Animal To Hominid

Part of the Evolving Series

(Notes Format only)

By Yildiz Ilkin

Library and Archives Canada

Library and Archives Canada Cataloguing in Publication A record for this title (2016) is available from Library & Archives Canada.

Publisher Publishing in process in Canada by Balboa Press for Blue Mountain Mist, Animal To Hominid undecided. A Division of Hay House 1663 Liberty Drive Bloomington IN 47403 www.balboapress.com 1 (877) 404-4847

Cautionary, Notice Of Liability

The information in this book is on an As-Is basis. While every precaution has been taken in the preparation of this publication, neither the author or its publisher shall have any liability to any person, corporate or entity with respect to the material content within the publication itself. All the materials presented are done to the author's paramount ability, in the fairest of academic standards, and researched well independently. However, due to its controversial material and to limit any liability; agreement clauses it has been labelled fictional.

Trademark

Kanata Means Wings is a registered trademark of Kanata Licensing Inc. USA. And licensed both in Canada and the United States Of America @2013.

When The World Changed

In a pitch black room curled nude, I am. All I hear is Yildiz you need to tell us what is going on. I play stupid at my victimization, I don't know. They answer back you do know very well. Your Ottoman notes, the world's first language is Turkish, clans, we all come from one linear source. These 70 documents, do you even know what you are writing? We have professors who are in academia for over a quarter of a century crying like babies at the embarrassment you just pulled. We had to make you into a theft, to save you and your children's lives. Do you understand us?

Members of your family, the 100 missing Van cats in one breeding facility in Turkey what a shame. Then changing the history of Canada's name, do you even have proof? This has become a national disaster for us.

Who are you I ask them, they don't answer? We are not CSIS this is all you need to know.

Listen I won't accept this I tell them we are governed by Law. You have to, they say. You can't have access to a lawyer right now and we also took control over your file in Geneva do you understand. You have been poisoned. We drain your frozen bowels each night, you will most probably lose your left eye once this microchip is removed. Big players like Russia and China are also involved, so many people are privately crying. Some cultures are very reserved and now you are linking them to Africa. Also, it is a horrible hazard for humanity when too much information flows out at once do you know that? It doesn't give others chances. If this wasn't bad, then you say shit like Jerusalem is Turkish and gave us the finger. Do you like pain, are you a nationalist they ask?

No nationalist, just a humanist, I graciously correct them. Jerusalem aka <u>Yerushaliyim has a proto-Turkic shamanic / Ottoman</u> derivative, the Turkish language itself is Europeanized and heavily altered after the republic's creation in the 1929 era. You know, like a prehistoric name. How can

I put it, when we think of people it is like Natives but on our side can you visualize this? This is the side I made into Landmass 1. Meaning the Anatolian or Asian side! As for the finger, it was my thumb not my finger, I am sorry.

Our global Indigenous people, wooo, a taboo word. I start screwing with them. Don't ever be ignorant or confused, there are 50,000 different prototypes of them. Our forefathers once did the same crap with the black community by making them one cease pool of "Negroid people only", living in their mega-humongous continent of Africa. After all this, I don't believe in anything academics produce anymore.

Who was your Ata **a**holes? I blurt out wanting to horribly die, it doesn't work! Yet feeling subconsciously excited amongst such darkness, for humanity, in unleashing our global heritage?

I reset, did you not read what I wrote I politely say? This is academic work a different type? I have sensory intelligence, I was doing labwork. In addition, these were my notes it had to be rewritten for my thesis and go through the protocol of academic review. They were stolen from me.

Let's get this straight they repeat in disbelief.

You are a Turkish nationalist with a PKK boyfriend, who got jealous and ran into a church after she found out he is having an affair with a younger Armenian girl. Then prays to be like a prophet to prevent omen. Yildiz what on earth does Canada tell NATO? Followed by the production of an astonishing record in terms of novel discoveries. Reports everyone thinks is stolen by the way, and with no PhD starts talking about divine and this religious book. It is mind boggling; did you reread your book?

The church helps us so much; do you think we can publish this crap. The President of the United States with his counsel around the oval table doesn't know whether to laugh or cry. Was this mental illness? Maybe you should covertly work for one of us they say. Was this worth this bad injury? Everyone is against you, millions were spent and I swear the source is not us.

I did not do this on purpose I say, in a low overture. No, honestly it wasn't like that. It is an irregularity but it has existed since the age of 4. You saw my hands. Listen funny stuff happens on this planet all the time, big deal! Let's all concentrate on the sciences, please.

The truth is I just had a dark moment in my life where several things at once occurred, even a vicious attack. It can happen to anyone; I got a hiccup. I am really just a good kid, a professional in her industry.

A hiccup? You just changed the Middle East and the planet they continued. These men and women spend years in school studying and reviewing each line. If we did not monitor you for months and see certain things we would never believe it.

I don't care! In the end I am fighting for the real Canada give it back. This planet is still in ape mode socially; I swear to you! I can now feel and understand all the prophets. Actually let's cut the hocus pocus I continue. In my opinion these historical religious figures were just regular people, made Chosen for having visionary views and for their irregularities.

Do humans really know religious history? That book that you took off the market glued together my sensory and a completely different story on religions? Why blacklist me, am I not under my Charter of Rights or better yet in the Borg?

You are an atheist; you get it don't you?

In the 21st century, I don't even like the word prophet! How passe-compose. Are you nervous that I stood up and modernized things? Let's call them now our Chosen people of earth that is better. The sad truth is, you can even include spiritual leaders, blessed saints, priests, monks or imams in this, anyone who spends all their life studying God and is humanly perfect, has to be blessed. Our religions that we have today was built by them, lets cut the crap. Our houses of God are littered globally; they totally need to change with technology. The word of God or meditative spirituality for people today should be an important component of stable existence.

In my case I know I am an irregular, so what! Or Chosen since it involves the church depends on your beliefs, and this is why. Prior to me all humans came from different sources. Not one single white Ivy League school would ever have the power to write this, nothing personal. Did you see my notes in Let's Evolve? Why, because it interferes with Ottoman, World War, Holocaust and our Slave history you see?

Know my sources I tell them instead of looking at me funny. Is this me or what? My examination of all the African stuff and human trait comparatives are also innately me. I was practically raised by the black community they mean a lot to me. My children being the foundation or sample study of that particular research document that was taken. Keep in mind, I simply went backward from landmass two, towards Africa. Using them as models, studying their irregular beauty for years. Hence making this Canadian more important than Nelson Mandela for toppling civil rights.

Then I went and gave the Turks their shamanic history with a discovery made on July 1st, 2013. I woke them up, by faxing the Ottawa consulate regarding what sections of my research was to come. And that the historic meaning of Kanata Ottawa, actually means "Wings Cover", a broader definition to just settlement.

Trademarked online and registered quickly hence making me more important than Ataturk. Yes, my confinement allowed me access rights to leave home so I ran and used a secure-coded fax hoping it delayed you blocking it. I know the mechanics of networks well; was that naughty or what? Listen this is our global heritage and Canadian history; I bluntly don't give two craps.

Now throw in the alterations of Charles Darwin and Mendel's theories in, this makes me more important than both men in our present day.

So I lost a few (politely said) marbles for producing all this, should I really continue? You are right, though it was too much of an unravel. I wont generate anymore and will donate some, I promise. But thank you or whoever, for only mildly injuring my unaware children. The sarcasm spills out.

This information is completely fabricated they respond back. Oh yeah, then why did you lock me down, like an animal I say. And what's with this red light in my left eye. Hey, this has to stop.

Our people made a mistake we are scrabbling to fix it. Yildiz we needed professionals. What does my credits in academia have to do with professionalism? They can bring out their own version if they like but don't forget the cc I tell them. You can never sell the sources of discoveries for the stunt you just pulled. Do I really need to teach these doctorates our plagiarism guidelines? Gee!

It gets worst they say, we had to hire actors and give it to the Americans, because of the Natives too. Here it comes the cynicism rolls through my head. The plastic pretty marketing people, I blurt out? No, you have no clue how bad it was. Absolutely no clue.

Forget media for a second he responds. We lost a few good men, Canadian CSIS officers, because they flipped it to the Turks and China. This was all the humour not realizing the validity in the intelligence material of the work you provided. We then ran to these countries and had to negotiate your own trademark globally to them as revenue to prevent further injury. An attempt to foil World War III with certain countries.

It's was the sensitivity of the language component that you wrote about. Threatened now became our First Nation communities all because of your work. Some smaller bands are literally missing. Remember Russia and China both have major weapons contracts with the Kurdish republic in Iraq. They don't give a crap about anything.

Then you bring out Africa too! Think about this hard Yildiz, regarding the proto-Turkic being the first language crap and don't be stupid. What this means for culture. You just slaughtered the planet, should we all learn Turkish now? Yes, why not I giggle? Cut that out, everyone in the major academic circles are calling you a nut job. There are a lot of racist people too. Then you make jokes mouthing off with your hand on your hip naked, how not only did they speak Turkish but how their great grand mama's was black too. You sure you don't want to do comedy instead; you looked cute though!

Comedy? Umm just so you revise yourselves Shamanic Turkic is <u>not Turkish</u>; Blueberry / Raspberry concept. I wrote a poem about it remember! Listen did this small amount of humour keep me alive? The things women have to do; I bluntly say to break barriers.

I wrote Blue Mountain Mist for a reason. Your academics know full well this broke a record; I stand by my work as being academic. Your agency should only be involved outside my world.

Did I break in anywhere physically like possibly England's old historical World War I records, was it the rich scheming or did the Vatican order this, after I privately spoke to the church about my irregularities?

This is not funny. Someone should be accountable. Have them relook at the outline of "Language of The Birds"? Be modern I say.

This gluing of years of hobby work happened after a prayer in a church, doesn't that tell you something. Plus, my children were born integrated beyond belief; I should work for United Nations. I am really the fairest one of them all, in bringing this out.

Furthermore, don't put these types of onuses on me, did I do anything to the Natives? Instead of injuring me I should be put into Buckingham palace I say. This is for balancing social culture, when the world was small between two large areas in historical Landmass 1. By you blocking the last three publications including Let's Evolve you have just changed the vortex of this earth. Do you feel guilty? Now I know why some people call you pigs?

The Queen is praying you will die Yildiz, and why the heavens did this to us as humans. She is the Head of State for Canada and is also heavily involved. We are trying to be fair. We needed many years to really review this, it does involve national security. Plus, be serious with us we are trained military.

Look we are going to be blunt, we had to steal all of this from you for another reason which is water. We have no clean fresh water globally; this will act as a catalyst in changing sea water to fresh water. Even your desertification notes, where do we begin? Academics distributed everything and started selling your book differently. They were tee off you either stole this or released this without research too early. The World and Canada will eventually forgive us. We are genuinely trying to help you now.

I glance at them oddly? The water was my idea and I have more on hydrology and environmental sustainability sell mine too, I only got my a** kicked. It is up to you I tell them. Technically its your omen of millions of people for not giving me a chance. Live with that! What if my one line changed something more major?

My webpage was my research centre for betterment. I am so badly injured I don't care what happens at this point. I am continuously delirious, I need rest. This planet was way too hard on me.

§

We want to tell you one thing you must have a horseshoe up your butt, inclusive of the US military they wanted to choke the crap out of you, but you are a struggling female.

<u>You are free to go</u>. Stay quiet regarding the blacklisted books they repeat, and count your blessings.

……. Are you serious! Can I know who he is, I say? No these families are very private but your DNA is very close to his. We know a lot of money had to have been flipped around, we are giving you another clue as to why certain things happened to you at the beginning.

He is linked to your true biography story of your now blacklisted Ottoman book. He got royally soft with you. He wanted you to know, he is proud of your cockiness; he knows full well you are touched by divine.

I will put what I can on Amazon Canada and I will ask the heavens that people in the future will evolve differently. Understanding the struggling and vulnerable; how I hope for a better place.

My aggression is going away. I am released, Legal is involved. Hence why you are reading this. Hopefully the year of the Lord, 2017 will be a better year. Oh what bravery it takes to publish all of this.

The following journal is one unit of my compromised and rehashed out notes, which were wrongly interpreted. By purchasing a copy, you are not necessarily agreeing to the content material you are simply supporting my initiative. One that caused me injury for bringing out our global shamanic history to that of highlighting our environment to land connection in terms of evolving.

The proper reviewed, academic edition will come out shortly.

The objective being to illustrate the many irregularities in our past especially in the realm of defining what is a human.

Dedicated to the many women in my life, their struggles and who helped raise me.

Just Notes

Author's Foreword

April Yildiz Ilkin has written this publication independently and of free will. So all the opinions, analysis and research are of her own. These notes will help in understanding what is our human network.

The study of early humans has always been an unknown area in its field of research, after all we don't have anything to really work with. Or do we? Academics have predominantly searched for clues deep in the different strata layers of earth to recreate as much as possible the mystery behind our prehistoric ancestors. Studying archaeological finds and attempting to guess-timate as I call it, this transitional period through their carbon-dated records. But this entire process has not been an easy task. In the end history has generally had a one sided explanation lost in the middle varying shades of grey, of what I call the black and white spectrum of the winner's account of our historical past.

Today technological advancements are accelerating human progression so much that humans can enter into multi-faceted domains of areas of interest making way and opening doors to our new information age. Which is how this intellectual ferment started. It glued together a collection of hobby notes, my field observations or independent academia as I call it and my commercial management knowledge over the years.

In addition, in one point of my life I took becoming an English teacher classes (ESL), to assist predominantly various Anatolian and Central Asian women who had no real formal reading and writing. Throughout this publication learning how to decode phonetics, dividing up of lexemes in words, understanding the different dialects and language structures, and grammatical formats of the major language families has aided me so much in my research.

Especially in noticing numerous familiarities with that of my birth language Turkish. But it didn't stop there, other than giving me language this region of the world also maintained social behaviours of intermarrying within clan family structures. This then became the interesting social evolutionary portion of my research. An attempt to see if we can amalgamate linguistic evolution with that of present day clan behaviours. The irony is my undergraduate degree at university was based predominantly off environmental studies. Always wondering whether our environments, clans and linguistic analysis, had an indistinguishable connection that academics had missed, my academic cross referencing of work began. Studying impacts from surroundings, social behaviours of clans and dissecting the unusual occurrence of why so many words seemed so familiar globally became the bulk of what is about to be presented. In addition, this is where the concept of linguistic, its study, as an actual auditory artefacts was born.

Interestingly Animal to Hominid came after my original series one, the notes for *"Let's Evolve"*. Since I felt when pushing human progress through the analysis of human social behaviours, which was the main thesis of Let's Evolve. A precedent was clearly missing. I realized a deeper look into early human development was absolutely necessary.

But can anthropology, our environments and linguistics which are three completely different areas of academia actually be merged together? Or can thinking outside the standard domain of research be a key to our past, even predating our oldest artefacts? The answer is absolutely. As a researcher the following publications can aide's academics and can be used as a tool. It combines the standards of what Harvard University indicates as intellect behind a different form of academia. It uses an individuals visual sensory, auditory and observational analysis to illustrates irregularities. Why is this important, because breaking perceptions plays a heavy role? This research paper is for people in more diverse specialized fields to further investigate. I would like to present the irregularities missed or overlooked of our past when technology was not that advanced historically and the flow of information was just not as abundant as it is today.

Before we do anything though I want to advertise this publication is written completely without prejudice and I also have no political mal-intent. Furthermore, as someone who is this passionate about environmentalism I can confidently speak on behalf of majority of us that most people who are that attached to their environments are also caring to what inhabits their surroundings. In general, these individuals have strong human and animal right inclinations. With the hardest part of this research being the ability to bridge together the division of the sensitivities of stigmatic politics with that of the academic world. In the final publication Let's Evolve my futuristic vision and a better world of a one that incorporate sustainability and no radical behaviours gets incorporated here.

As for myself, my parents conceived me in a small village in the mountains of Ararat, born in Istanbul and I blame the spirits of our ancestors and these groups of mountains for my greenness. I have travelled the globe extensively and come from a multi-ethnic, mix religious, and very vibrantly cultural background. Just this type of multiculturalism is a gift in itself because it assists me in helping or teaching readers what I see as we attempt to cross reference our cultural connections.

Lastly this is a reminder to all humans one of the biggest weakness we have in pushing progress are the spatial analysis of time merged with heated politics. An understanding that we are merely a single component of time, a milli-second, in this vast human timeline that has existed. The objective of Animal to Hominid is to expand on this. That for the average person a million years is not much different than a 1000, but the problem it presents is understanding the real component of the numerical values of zero's in our history. And attempting to quantifying in the mind what I call "*the visual aspect of time*". This research from start to finish has another component to it and that is the awareness of time and its impacts on human incremental development. Animal to Hominid starts this process of understanding. Defined as what was our historical views on progress, how has our environments help us mould it and how has it grown for each year that has passed.

We have come so far in this time period, this era, known as the 21st Century. It should be reminded that the spirits, the chi of our many indigenous ancestors that have come and gone have left us something for our advancement and comforts today. I call what they left behind a reminder of their existence, humorously said - "X *marks their spot*" our spirits are still here to guide us into the future.

Enjoy! Yildiz Ilkin

Study the past if you would like to define the future...

What are tools? Tools are specialized icons to assist us in learning. They are there to help recognize certain human patterns and pinpoint their environments role. But before we jump ahead in our developmental journey I have divided this publication in the following categories to give a brief overview.

1. *Tools and Icons*
2. *Criteria for Research*
3. *Content Material*
4. *Visual Content / Educational*
5. *First Dictionary Index*

All the associated icons used throughout this book are as follows:

1. Visualization Techniques & Methodology

Throughout this research guide I have put in a series of visualization exercises, the methodology is quite simple, with my guidance pull back from the computer, close your eyes and visualize. It will be our imaginary method of going back in time and understanding the topics discussed. Why is this tool so vital? It helps us focuses on two components one is the comparative theories presented throughout this publication, usually always using the quantitative measure of time, the other is the academic component of the research. Now let's remember to visualize all of the highlighted asterisks throughout this guide! It will use reasoning and / or the cognitive portion of the human brain in understanding what went on historically.

	Visualization Icon

2. The Importance of Academia

Incorporating scientific or scholarly research on a variety of topics throughout the publication should act as our support guide in helping us develop. Let's take simplest example the "*Animal Kingdom Comparative*". This was scientific research that was done to show our DNA connection with animals. In this particular case why is the study of animal behaviours so critical in understanding human activities and evolution, because according to the Smithsonian book on Humans (Coordinators, 2004) by deciphering genetic codes, scientist are able to understand the links between species.

This shared DNA suggests our shared ancestry. Simplified it means other than what Darwin wrote regarding species evolutionary development and what has historically been adamantly rejected by millions due to religious beliefs, scientist are now telling us there is actually a genetic shared ancestry with that of animals. To understand evolution on a larger platform lets listen or take seriously what our scientist are telling us. These are valuable people who spend years in very specific domains and are experts in their fields of knowledge.

	Academic Related Information Or Statistic Icon

3. The Highlight

In the following chapters when you see this star-symbol, like the one right below, it is just indicative that it is associated with an interesting fact. Interesting facts are there to shed light on what has happened in the past or to highlight certain pertinent information.

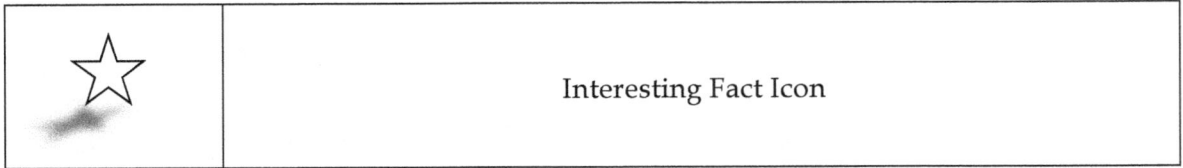

☆	Interesting Fact Icon

4. The Factor Of Time

What is time? Time is a measurement module that needs to be emphasized continuously due to its lapse or understanding in the human psyche regarding an actual quantifiable measure. In our case when we look at time it is the duration of period intended between two events to that of our human timeline history.

Many of our explosive regions of the world and how they have developed that way, are so new in retrospect to the civilizations that have existed prior that only a "forced realization or highlighting" on the topic of time can help us focus on the sheer number of epoch periods that have gone by. The "emphasis" for example that 10,000 years may be relatively old for the average person but is so new in our human time line, is what we are attempting to illustrate. Now let's remember that time is a component the average human has extreme difficulty grasping. When you see this icon, try to grasp the actuality of the age that is being presented.

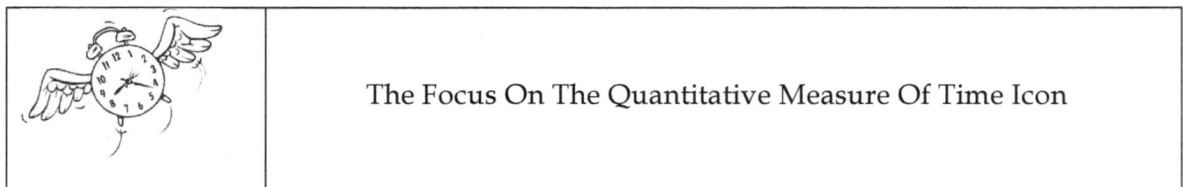

(winged clock)	The Focus On The Quantitative Measure Of Time Icon

5. Items In Italics

Terminology, definitions and any Items found italicized can be referenced with the index section under glossary, at the front of the publication. There purpose is to present critical explanations in a per verbatim, collective manner.

ITALIC "Words"	Italicized Definitions

References

1. *Smithsonian book on Humans* (Coordinators, 2004)

For every living organism out there, life started from a single cell. I called this the transfers of life, or the step by step of how it all began.

When we look at our environments today the average person does not put too much thought into it as we scroll through our surroundings, which in majority is made from piles of well groomed concrete. But has the boom of urbanization reached a plateau in terms of human growth and development. I believe the answer is yes and a deeper look into our environment's role over time will illustrate this. In the end no matter which part of the globe humans live at, our goal should always reflect a harmony with the environment we live in. This environment is more simply put the world around us. Based on two back to back publication, both series have different themes but they are intricately built on an understanding of the role of our environment, its associated outcomes on human behaviours, its exploitation and the need to preserve it.

The inspiration behind Animal to Hominid begins here. It attempts to demonstrate how it all began. It summarizes a detailed account of what the animal kingdom and surrounding landscapes have contributed in the maturity of early humans. It is a foundation, a base, a precedent which is needed to illustrate how socially we came to be. Why is this critical because a peak into our past will help us understand our role as humans in the habitats we live in today. It demonstrates the critical importance of our environments, its every contribution, and why her protection should be carpe diem of every minute we exist. As one indigenous tribe told me "it is not us that will destroy earth, it is your ways". Is the paradox behind the publication. How our attempt to be advanced in one way failed so horribly in the next, by not understanding the very basics of life. And our connection to the world we live in today.

The second publication detours into a whole new different dimension. "Let Evolve" part two highlights our primitive behaviours, done in the name of social self preservation. Better defined as what humans have done in the name of survival. The basic economics of limited resources to that of an ever growing population and the study of social behaviours behind the absolute atrocities that have existed out there. The art of war techniques of our past global history will slowly start to unravel in illustrating how still animal, we really are. The aim is awareness and why a look at the past may be the answer in resolving the future. It emphasized our interconnectivity and how we have evolved technologically but still require primitive behaviours to be manageable in our own environments, even sometimes destructively.

The sad irony is from the buzzing of a honey bee which pollinates and aides the $200 billion dollar food industry; to the visual of landscape that gave us the transfer of language from animals to early human sounds, we are all interconnected. Everything is weaved delicately into the very fabric of our existence, down to our very last breath. For thousands of years when there were no shovels our ancestors used birds and animals to clean up the carcasses of dead, a form of natural recycling. Today vultures are becoming extinct. Like many other animals that are disappearing, the circle of life has been halted by human's confusion of what it really means to be advance?

One key area in environmentalism therefore is our role in monitoring sustainability of our landscapes. Surroundings are always at the forefront of any area in the field of green an environmentalist will branch off into. This of course is always an area of study pertaining to one or more categories under the "*environmentalism umbrella*", as I call it. A larger range if we even merge social behaviour to that of our environments of what the average person see as what is historical environmentalism as illustrated in the diagram below.

Environmentalism (Table A)	Progressive Environmentalism (Table B)
• Global warming • Chemicals • Pesticides • Trash Climatic monitoring	• Study of historical social behaviours • Environment & Linguistic development from land • Land generation to that of • human physical traits • Collective reasoning through understanding the above. <u>In Series Two</u> • Landscape & Mental health • Historical Impacts Generated by Environments Recommendations & Social betterment

Known Environmental	Hidden subtopics
Management of natural resources	Mental Health
Climatic Problems	Women's Issues
Impacts of Concrete Our artificial settings	Genetics Our dissociation from innate requirements for greenery (Our animal self recognition)
The Tumour of Desertification	Poverty resolution to that of human violence
Islands of Trash	Human cognitive development
Biodiversity & Ecosystems balancing	Psychology and social impacts
chemicals	Ethics
Ozone and heat	Business

But other than the list provided in Table A, when we think of our environment are there other areas never really charted before? Could a landscape so green and living with a multitude of species actually have given human beings something more than just the physical objects of exploitation we know of today? What about the contributions of our environments to early humans in terms of visual and auditory development to that of thought, or that of actual physical appearance? How did language or social behaviours develop?

Animal to Hominid is the step by step start in analysing what we did and how lack of awareness on behalf of our forefathers has generated such damning historical outcomes in terms of what it means to not only be progressive but also balanced in our own ecosystems.

research Once upon a time 350 billion years ago bacteria in the shores of Australia which was then attached to South Africa brought life to this planet. Antarctic emerging upwards broke the Australian sub continent and actually generated prehistoric life in south Africa. //Deduced based off the age of the most destruction and traffic in our oldest deserts Namibia regions. This is where life as we know of today began. From this point onwards we start to piece together our link to the animal kingdom and our environments.

Self Replication

The start of life will be examined as the first bacteria started replicating, its engrained behaviours, followed by the creation of all living creatures of earth. How the concept of self replicating, created the first human species of earth. How a group of inter-mating first clans of earth develop to the top of the food chain. Our humble origins commencing out of Africa and during their migration patterns how our environments shaped us. These are called our first incremental steps of hominid development. This process was over the course of 3-4 million years and was no doubt painfully slow. This publication gives humans an understanding of why understanding self-replication and interconnectivity today is so important.

Nomadic Migrations

The hypothesis is as some of our largest prehistoric dinosaurs destroyed the oldest true desert on earth Namibia approximately 70-80 million years ago, with our five extinction periods our start of smaller animal transitions began. During periods of harsh climatic variations, the damages in and around Namibia desertification years later contributed to groups of our earliest human predecessors starting their treks out of Africa looking for food and the start of our earliest dialogues.

Our scientist have found the oldest intact human fossil to date in Kenya, so how can I be so sure that our earliest humans actually started their journeys a few thousand kilometres south of Kenya near South Africa. When it comes to history no one can be sure, social ethics and DNA will be discussed in later chapters however this publication will attempt to illustrate the hypothesis and the origins of the first voyage through a interdisciplinary study of various factors, which are unconventional to book academics per say. Starting with using rationalization as we attempt to visualize the "Man who Walk" theory to that of integration over time. Doing observational work in attempting to catalogue physical features globally to that of the study of first clans. Lastly bringing out our green sensory by understanding nomadic movement of our hunter gathers searching the globe for clues for the locations with the worst desertification damages. Our goal being to evidence our earliest hominid colonies.

These factors are all indicators for the origins of our earliest animals of earth, inclusive of early humans, which I call the regions with the heaviest traffic for the start of life. Damages to earth aides academics in providing humans with clues of outflow of the origins of life and routes of first nomadic patterns. Or if we are date sensitive some may call it the movement of our earliest animal migrations.

The Birth of Clans

What are clans? Clans are our earliest groups of same family members, or descent. Today clans are still found in a few locations globally, predominantly more so in what I have labelled as the first civilizations of the world. A more specific example is our case study the region in and around southern Turkey where they are commonly known with their alternative name as Ashirets (*pronounced ah.si.rets*). No different than our earliest forefathers comprised of smaller groups out of Africa. These groups of families limited their marital unions only amongst themselves as intermarriages or unions between close family members. They also have a holistic nature for self preservation and their responses to their immediate surroundings are very tight knit amongst themselves.

Introduction of Shamanic Proto-Turkic

Language all started in the animal kingdom. Animals have always had constant dialogue, first between same category of species themselves and then the inherent art of mimicking other species. This publication will demonstrate how our environments have contributed to the languages we have today. Descriptive language analysis from Africa to the American continent comparative aides researchers in understanding our earliest connectivity. We will be using the study of language extensively to not only give us clues to our past but the visuals necessary in understanding what happened in our earliest ancestors ways of life.

Regional markings also known as Birtur will be introduce as a new classification of language also known as the period *"when there was no written"*. It will be our guide to a different time period. Its core makeup is in the format of Pictionary descriptive, or X marks the spot visuals. Pictionary descriptives to our past is exactly what is needed to help us advance. This is our missing link which evolved us from the animal kingdom. Words associated with early humans give us the data necessary to record our historical past.

Greece for example contributed so much in terms of enlightenment historically. We have come this far by their mere highlighting of philosophy, law and democracy etc. This publication is no different for its understanding our linguistic evolution and its birth from the animal kingdom. It grounds humans by moving us away from the radicalization of language, it illustrates our interconnectivity as humans, and how territorial controls began. It shows the before and after look of primitive humans to that of more advanced societies. Respectively being the *Animal to Hominid versus Let Evolve* comparative. In actuality it is exactly this combination which contributes to an entire process of changing human perceptions. Isn't being aware and consciousness a form of education after all?

The notion that the moulding of environments to human behaviour is what I call our first key area of awareness to advancing. It also show how much animals and landscapes contributed to the very existence of humans today, better defined as the transfer of life.

No Mal- Intent

This is where the dissection of academics to that of politics is really needed. Since part of this publication pertains to linguistic, and language has evolved over the thousands of years to be very radically associated with a territory. I would like to highlight that I have no political affiliations and I have attempted to be as humanely accurate as possible with all documented sources. The content material, observational recording and information in the index in the back of the publication will be my evidence that something was clearly miss in our historical timeline. Now let's open up many other areas of research. We will be using descriptive language and the study of clans to understand our past, to that of our contributions from environments and why we have done what we have done historically.

References

1. *BBC Earth live*

2. *Encyclopaedia of Earth*

3. *National geography Genographic project. webpage www.genographic.natiornalgeographic.com*

Chapter Two - Criteria's for Research

When it comes to history there should be an understanding amongst countries. Researching, validating and recording sequence of events, are our only measures of human progress.

Animal to Hominid criteria's for research especially in the areas of linguistics will be our start in trying to organize, assess, collect and decode historical material. This section is necessary to create some type of standardization in order to evidence the study that is being published. For each topic area in the following chapters, a small square box at the start will provide readers an insight of how this material was pooled together. Why is this so important because we are attempting to evidence the *"unknown or unwritten portion of our history"*. With the whole objective of this publication being the use of non-conventional methods regarding academic researching. The concept more basically referred to as having humanity thinking outside the box in retrieving historical data.

Since this publication pertains to very sensitive topics like what is skin colouring, hominid's linguistic, birth place of humans, defining what is a human or linguistic evolution, our goal is to limit errors. It provides readers with where the main challenges came from, the irregularities that exist and open up many other areas of research. As a researcher the critical sections have all been given a summary of how this publication material has been derived.

Anthropology Section

Introduction of clans social behaviours in their environments, clan unravelling to matrilineal marriage formats, human mimicking tendencies etc. The sections on clans is easier to account because there are no real criteria's like the research on the linguistic portion. Instead it was based off reporting and researches of mainly two different social groups and their ideologies. Members of communities belonging to Panamanian Indigenous and that of settlement in Southern Turkey. Furthermore having travelled and visited clan based communities, and interviewed numerous people on clans in the Nigerian community. I merely presented findings in this section. The most critical component of clans are the norm in some regions regarding what I see as self preservation and their human social behaviours of continuous preference of intermarriages amongst their own family members. A look into areas of antiquity like Africa, the Indus Valley and Anatolian regions provide a glimpse of our earliest animal hominids transformation to human, and how life actually developed. It will be our illustration that humans are one concept in helping to evolve.

Bitur Classification Criteria's

Later on in the chapter we will introduce Birtur which is a language classification I have labelled based on the era of "*Pictionary Descriptive*" and also known as the language of our hominids. It is a general term referencing Shamanic sound classifications based primarily on descriptive visuals or markings of regions. Birtur is what our first humans saw as they attempted to convert certain associations vocally. These sounds and single syllable words are a direct results of the impacts of surroundings and its association to the development of language. A unique classification that I summarize as follows:

1. Topographical environmental based sound to that of single syllable Word Conversion;

2. The start of Pictionary Visuals and identifying; and

3. Lastly transfer to dialogues, full regional marking within early human perspective territories.

How can we standardize or conclude merging the following list above with that of researching historical words of regions? By attempting to break it down into five general requirements. Pictionary Descriptives therefore is the study of sound and word narrations based off certain global patterns collected. The question being was there a consistency in these hominids / humans in their earliest dialogues. The answer is yes there are very strong patterns that would classify this research slightly different than the study of the etymology of words for this reason. If we think about the National Geography's Genographic project for example we think about the movement of DNA. Movement of linguistic is no different, which historically was never confined to a particular region.

Etymology is where the word was first seen in terms of origin but may not be where it originated from. Later on proposing to use DNA hand in hand with linguistic research is where Pictionary descriptives becomes critical in terms of academic study. Especially in areas of DNA to linguistic cross-referencing and studying the movement of language families.

Unlike what etymology of words mean there is a clear distinction. Pictionary Descriptives are visual expressions or attachments to that of confirmed areas, historical words of importance derived within a region, territory nicknames or that of an animate objects used locally in regions of antiquity today. Prior to "overlay" meaning when one language dies and another begins. Pictionary words, its earliest recorded proto-Turkic format is related to human Shamanic history and can also be expressive in nature. What we are attempting to evidence is that it belongs to our earliest ancestry, while etymology is defined as a word that is newer in human timeline and most probably exists in a written format somewhere that was recorded?

Throughout the index guide in the back, the word patterns and descriptives used will have to pass the five criteria's to be regarded as belonging to Pictionary Descriptives language classification. Please have a look below at the criteria's created in helping decode and decipher our unwritten portion of history.

FACTORS	DEFINITIONS
1- Does the word correspond to its periodical dates?	The dates we will be referencing are our Shamanic history at approximately 1.5 million years old to 10,000 BC. This is known as the "*stagnant period of life*" where dialogues was linguistically stable in retrospect to its hominid and human population numbers.
2- Is there substantial evidence?	Are there other closer settlement where the words can match on a micro (local) or macro (neighbouring country) scale? We are looking for about 450-500 pockets of nomadic settlements dispersed globally. These are called individual territories to respective clan families. How we derived at the 450- 500 mark will be explain in the next chapters.

3- Are there consistent patterns regarding choices for topographical regions?	Are the location consistent to other areas where these types of words exist, research has found them predominantly near fresh water tributaries outlets by coastal ways, mountains and / or islands located near land? I call this period, when inside rain forested areas were not a favoured or a safer choice for settlement and our earliest humans had not discovered aqueducts or water tables underground. A good example of a hominid colonies for example are territories or regions near fresh tributary outlets.
4- Does it fall into any recognizable word classification?	Some of these are phonetically decoded descriptive words. Can we see slight modifications in words if we break it up lexeme wise? Are there any changes done by the local languages of the region today? Adding in a first letter, modifying the ending, changing the word koy (means village) to sky (Russian) ending. A good example are all the suffix addition. Taking Karakawa (dark head clan) and making it Spanish in format Carakawaque a region now found on a Spanish map - where the k and the c are the same phonetically and the que (aslo found in French) is merely a Latin based suffix add on.

5- What's its definition?	Defining the item or decoding it. Does it fit the criteria as a descriptive region, or a descriptive action, based on its probable date?

Please find below one example from both an action descriptive or a regional descriptive below:

ACTION Descriptive Word

(a) The case study of the word KAR-BUZ (Snow-Ice) from India / Pakistan.

So I walk into an ethnic store and there is this odd shaped orange fruit, with the label Karbuz from Pakistan, I start investigating because Karbuz decoded phonetically means snow and ice. Why on earth would Pakistan, with a completely different language have a word that means snow and ice. My investigator hat comes on and I start researching. If we understand Birtur's criteria it sheds a light to our past.

This is when the world had no refrigeration, a striking similarity comes to mind. A flash back in time and merging cultural knowledge to a remote village in Turkey where locals still bury vegetables as their natural refrigeration process, under piles of snow. The beauty about first civilizations is that very little has changed, other than some infusion of technology from the outside world.

Karbuz today is linguistically transferred over in Anatolia to the word karpuz (watermelon) with zero affiliation to what our indigenous forefathers did with the actual odd looking orange fruit. Karbuz therefore is a word I would classify as Pictionary descriptive in this case it is descriptive of "an act of doing something" of our earliest ancestors.

Let's try the check and procedures in place, called the five criteria's to make sure:

1. **Does the word correspond to its periodical dates?** The analogy is many animals hide food underground. This act is descriptive to a period when there was no refrigeration. It's the action of digging into snow and breaking ice to preserve fruit, which is researched to be centuries old. The word itself broken up, Kar (snow) and buz (ice), is a behavioural act of doing something in snow ice and is indicative of its age. Today centuries later Pakistan is not Turkic where the fruit originates from yet the reminiscence of linguistics and its historical meaning remains.

   So yes it fits timeline next √

2. **Is there substantial evidence?** This word comes from an area historically known as Hindi Kush corridor / territory - Hindi Kush touches Pakistan and historically belonged to our Shamanic forefathers. How do I know, Hindi Kush translated literally means tribes of landing bird clans in proto-Turkic.

   Yes it fits next √

3. **Are there consistent patterns regarding choices for topographical regions?** I asked a local regarding this fruits and they said it is located in a mountainous region and the higher altitudes is where it was buried. A mountain, with its water tributaries to sustain life. Yes it fits next √

4. **Does it fall into any recognizable word classification?** The first dictionary of our earliest ancestors was very basic composed of adjoining single syllables, this word broken up is a complete sentence and it means the act of putting into snow ice. Yes it fits next √

5. **What's its definition?** I classify this as a good example of an action belonging to a Birtur Classification era and part of anthropological research, a descriptive meaning of what our earliest ancestors did. The language is no longer Turkic, the word is basic and forms an entire sentence, and it gives a clear visual to our past.

Yes we conclude √

Here is another example this time using a criteria for a regional descriptive.

REGIONAL Descriptive

The case study of the word Dokkum, in the Netherlands.

Here is where it gets a little complicated, as I spend hours looking at an atlas I am attempting to categorize these descriptive patterns of words left behind and this is why I have used this particular region as a good example. It also illustrates some of the difficulty I face. Keeping in mind I am not working with technology for carbon dated artefacts. I am just swimming through mounds of word and gluing patterns together meanings while I investigate. It requires more than a guess of the actual word meaning itself. It requires the ear of knowing the base sounds of several languages, or dialects, understanding culture and the breaking up of words.

For example If we read this word in modern Turkish today Dokkum in the Netherlands, sounds identical to the word dökum, which means to cast metal. The problem is the act of casting metal does not fit into Birtur criteria's for age, since I believe these clans were way to primitive in the act of casting metal.

So I go to round two of investigating which is to listen to someone Dutch pronounce their regional word Dokkum. Out it comes the hidden T in their pronunciation which sounds equivalent to the standard Turkic family of words Tokum. Like a piece missing out of a jigsaw puzzle it fits perfectly into our general patterns.

Tokum means *the act of being full*, now why would the Netherlands, near a water tributary possibly have a word that means full in Shamanic proto-Turkic? As a little introductory clue into later chapters the concept of being full was the Shaman's ways of disposing of decease, via birds and animals. Understanding that this was a regional point where feeding human remains to birds or animals was common in periods when shovels did not exist for early humans. Could this be completely coincidental absolutely but like illustrated in the back of the book the patterns presented are way to organized to that of primitive humans way of life. In addition they match closely to Eskimos and North American Natives format of early burials. Native study comparatives to Shamanic behaviours in Landmass 1 are an ideal casestudy in assisting us for research.

Using the five criteria's listed above lets dissect the word Dokkum just to make sure we limit errors.

a) **Does the word correspond to its period?** Birtur dated era is anywhere between 1.5 million years old to 10,000 BC (prior to any major language changes and overlay something that will be introduced in the next chapter), no different than North America's indigenous (in Landmass 2) it was the feeding of animals when there was no shovels a common method of disposal.

Yes it fits next √

b) **Is there substantial evidence**? The Netherlands are close to Finland which has a confirmed 2% of Turkic in its language. Not Anatolian Turkish, but rather Indigenous people's Turkic. This is the same analogy as when Judaism was found in Ethiopia, in reference to the variation of types of people with a common denominator. Yes it fits next √

c) **Does it fall into recognizable word classification?** By breaking Dokkum into single syllable you get *Dok*.<u>kum</u>, Kum or Dök or Tok these all belong to the proto Turkic family of languages even today. Understanding its age, these single syllables are found commonly throughout central Asia and Anatolia. All separated they are common words and can be used interchangeable in any agglutinative format. Yes it fits next. √

d) **What's its definition?** Does it fit its criteria as a descriptive region, yes because it means the act of being full and we have similar words which also means full in North America's. The regional index will list them. Yes we conclude √

CASESTUDY The word Kalamata, Southern Greece

The separation of Birtur related words. Politics unknowingly has had a profound impact on human progress and development. This is why we are attempting diligently to separate shamanic proto-Turkic from that of Anatolian influences in the regions around the modern republic of Turkey. Our criteria's were setup to catalogue, no different than cataloguing DNA, and look into the world of our earliest human linguistic out of Africa. The analysis of these words therefore is a form of study that needs protection and can be a critical tool for anthropological research.

Let's begin with understanding time. There is thousands of years separation between Birtur (a classification of shamanic dialogue which are inclusive of hominids) and that of Anatolian, Turkic influences of a region. Ironically similar sounding words of Turkic or its proto format origins may exist but this may not necessarily fit into Birtur's five general classification criteria and could be uncorrelated or part of an empire like the Mongol, Byzantium or Ottoman for example that always had pockets of Turkic language in it. If we were to analyse Kalamata a regional word found in Greece, we can see how this does not fits into our five criteria's:

1) **Does the word correspond to its period?** Dates based on research of Kalamata is periodically newer since broken up they are also co-shared linguistically with Ottoman Farsi/ Arabic.

In the chronology of linguistics section we also know if Ottoman language is involved this is completely out the shamanic era linguistic so, No it does not fit next. (X)

If you have even a base understanding of historical empires in the Anatolian region this will be enough to know that during these periods there existed many dynasties, integration and intermarriages. Kalamata (Kalam. ata) broken up means my lasting leader or an adjoining name of a leader, possibly an administrative beyliks' territory . How can I be so sure it is not a Birtur classification, if we look for other patterns of leader derived words, in closer regional points and merge it with historical research it glues together.

We found India has Kalikata (Kalik ata) Calcutta's old name and Bulgaria has Kulata (Kul ata). Having links or influences of different rulings of empires should not be confused of the period when language was first developed out of Africa. Are you starting to see the difference between Kalamata in Greece to a word possibly 40,000 years old (+) like Dokkum in Netherlands. One has to do with settlements, empires and regional influences the other strictly falls under what I call indigenous shamanic related timeline.

This is also where Turkic academics worldwide need to step in since only those who understand the language should be able easily differentiate a Shamanic related word to that of a settlement via auditory analysis. Since there are very clear differentiating sounds that both era words will project. Meaning if you have a Turkic language background, and it's highlighted, you will immediately hear the difference in the variation of words, this is why an audio will soon follow to help evidence our shamanic history.

2) **Does it fall into recognizable word classification?** This is related to some type of historical rule of leaders based on other comparative regional words. Just the pronunciation of the word

Kalam 2 syllables more eloquent to Tok (1 syllable) that is primitive and vocally hard gives us another clue. No it does not fit next. (X)

Investigative Portion

Where the word is located, its age, one is Anatolian settlements the other is hunter-gather immediately removes the regional marking Kalamata out of Shamanic era of dialogues. This word found in Greece therefore has failed the test of being shamanic related. Does this make sense? My answer is definitively since Greece provided advancements to the world and had abandoned shamanism-type lifestyles centuries prior. Minus of course the territorial rule of various empires, sultans and influences in the regions.

The attempt for us to categorize section is designed to aid readers in understanding how this research was derived. With the focus being in countries where Turkic is not spoken today. The numbers and the spread of communities that do speak Turkish or indigenous Turkic are already evident. This simply shows that a form of nomadic proto-Turkic existed pre-global population explosions. Professor Alfred Toth had asked a question are all agglutinative languages related, the answer is definitely. My response is changes in words over the course of thousands of years would never change underlying sounds from our environments of first clan linguistics. I call this the format of naturally and incrementally built dialogues over millions of years.

References

1. *Ethnology book of languages*
2. *world of languages*
3. *native encyclopaedia*
4. *list read*

Chapter Three - Error Limitations & Challenges

An error only becomes an error if it does not provoke thought and communication?

When research is presented and there are no ways of substantiating the findings to a 100% accuracy how can we create a universal approach to try to minimize errors. The key challenge then becomes how do we research something that may have occurred a million or two, years ago and provide substantial evidence to prevent getting a backlash from academics or the general public?

In the process of developing "*Animal to Hominid*", the biggest challenges that have occurred are not the investigative portion of the research but rather the following:

Challenges

1. In highlighting language there is a direct association to that of a national flag and carved out borders. These engrained thoughts in humans are a direct result of domination of land and the need for self preservation in humans with that of population increases. Especially in more recent years, I feel one can also draw a parallel to population increases with that of globally lacking sustainability initiatives in humans. In some cases these thoughts of fanaticism are so severe it has moved us away from academics and is focused on the radicalization of language with that of cultural identity in humans. Shamanism earliest links to a small group of roaming hunter-gathers found worldwide, who started dialogues in the earliest proto-Turkic format poses a challenge.

2. Dealing with the anthropology portion and having to explain to the world we are actually very connected to females who were not Caucasian in visual appearances touches race relations and poses a challenge.

3. Explaining the descriptive word meanings to an audience that does not speak the language present a serious challenge. Can we trust this author that is associated with a hotspot country to present these findings accurately?

4. Bringing out 2 million years of missing history from our ancestry and how everything was ignored, hidden or destroyed previously due to various involvements of empires, religious downplay of pagan behaviour and powerful countries own agenda's. Can bring out centuries old ill feelings. Especially if we merge Europe's colonialism ideologies, World War I & II politics, Ottoman breakup aftermath and Operation Torch in and around the regions. Meaning politics historically has gotten in the way of what is actual academia. These engrained thoughts from 100 years ago poses serious challenges.

5. Change is the only constant of this planet. However any concept of change especially those related to territory and its developed human extremities can bring out detrimental emotional wounds to some of the communities involved. If we do a comparison to the animal kingdom it may even bring out our competitive nature, for example who and what colours crossed the Bering first? This simply put is part of our thousands of years of human intrinsic social behaviours and it does pose many challenges.

6. More so in the second series, defined as what you mean prior to the major religions our earliest ancestors actually spoke, had culture and beliefs in spirits? May interfere with religion and pose an association to paganism that won't be religious friendly. This is especially true for people who have hard core religious beliefs in God and that of the Adam and Eve concept. This poses different type of challenges.

7. Then there is the tiptoeing through the possible outcomes of venomous politics, racial bigotry and centuries of engrained distinctions amongst humans. Distinction in this publication is defined as early groups of peoples ideologies *"the concept that I have to be distinct from you, for our own clan's safety"* is centuries old. Sensitive regions like Middle East, Anatolia or the events pertaining to the Armenian genocide poses challenges in understanding what really is Shamanism? Followed by overlay, quantitative measure of time and what is social patenting in humans. Poses challenges.

8. Broken down sounds through humans making considerable progress in dialogues over time and taken from the incremental building of Shamanic Proto-Turkic gradually went through the many language families. Take for example the word Kalamata in the previous chapter. Over hundreds of thousands of years especially with a shift from nomadic to permanent settlements and many languages later these sound to word conversions created different meanings. There is no question the age of Shamanic dialogue and how words transferred over to become co-share words in several languages do pose challenges. Greeks may say "the Beylik or Ata example is a complete coincidence, Kalamata actually means this today and is now associated with Greek nationalism" please can you stop lying. Attempting to educate, erasing borders and attempting to catalogue Shamanic sounds like the word Ata, al, ala, kal, kala found globally do pose a challenge.

9. Proving to academics there is an error which was very politically driven at the time may interfere with numerous Universities research. One example is the conception that we don't have to only analyze monkeys noises anymore in understanding dialogues (like years of research by present day university of xxxxx is doing...... (find university) we will evidence that we have 2 million years of certain words in their base auditory format and analysis in the language of Shamanic proto-Turkic itself which is worldwide. Pictionary descriptive which will do way better in terms of merging the gap between hominid to human dialogues. University of .... may be devastated by the research presented, this is where If we think creatively, effective today our attempt to catalogue hominid / early human sounds can actually be merged with that of monkey noises to get an astonishing result of human linguistic time-lining. Unfortunately this is only one example and some academics may still feel this is a form of undercutting of their past work, and this may pose challenges.

10. Lastly the ability as humans to not be able to disassociate our earliest sounds or words, which is part of our global shamanic history, from that of a hotspot country known as the modern republic of Turkey may create adverse reactions. It is exactly these stereotypes, lack of knowledge of the multi-religious, physically different indigenous Turkic speaking people out there and the association of Anatolian Turks to Islam that could opens doors to explosive reactions internationally. The concept that an Uyghur Turk is the same as what cross over into the America's 50,0000 years ago to that of Anatolian Turks or the negatively presented statement "what we are all Turkish now?" poses serious challenges.

Limiting Errors

However, before any spectacles of fireworks or fierce rebuttals may occur, there were other sources used as well behind the research presented. The authors own background, the holistic incorporation of reference sources, including National Geographies Genographic project will also aid in limiting errors:

> ESL training and attempting to do auditory separation of words within the Turkic family of languages themselves. What is Shaman versus what are settlement words, to that of a base of Canadian Native linguistic that are co-shared with central Asian Turkic linguistics. How the first visuals of our surrounding gave language. Example Aristotle would never be Aristotle if he didn't have the existing animate objects of his surroundings and the associated sound attachments that had been initially established, to develop him further notion;

> Academic Reference sources especially in the area of sociology and environmentalism to glue together human development to that of the critical component of surroundings.

> Traveling and recording. Presentation of what are first clans and observational behaviours regarding different regions, customs and clues that started our humble beginnings;

- The <u>visual of the DNA migration map</u> of our earliest human journey, migration routes beautifully laid out by the Genographic Project tells us the planet is not just majorly integrated but also has just changed.

- The America's I feel are an independent land mass broken off approximately 8000 years ago, working backwards from Landmass 2 to Landmass 1 and vice versa, then referencing <u>Native resource books </u>gives us a great start of early history comparatives with that of Africa / Anatolian or Indus Valley customs.

- Lastly evidencing language order with the help of <u>linguistic books</u>. Gives us the following information: The known age of some languages, which cultures had alphabets who didn't, who piggy back off others for the start of written. What language families takes longer to learn, whose are more natural in phonetics? How did our exploitation of our environments, possibly give way to the very complex languages we have today?

In the end all we want to do is rationalize how we started. We want to go back in time and understand early human psyche in what I call their *"natural habitat"* and what created certain behaviours or developments. Our goal being to limit errors because of the different type of academic work involved since it concerns the investigative *"unwritten portion"* of our history.

A researcher who merged certain observations in regions of antiquity and is ESL trained in understanding various Turkic groups may be able to dissect words but will run into several limitations compared to a multifaceted team of actual Academics who can spends years researching the history, incremental build our environment / earliest humans and meanings of hundreds of regions.

Basically this work is an introduction with a lot of lobbying under the legal clause, of democratic freedom of speech of our hidden past. Trying to present to ivy league schools what

I call a major academic irregularity, even in the anthropology section, the two side problem as a result of the historical impacts of slavery and / or ignorance with the development of different humans of earth. Which may have left academia vulnerable and/ or not questioning the works of the past departments they were trained in.

References

1. *Ethnology book of languages*
2. *world of languages*
3. *native encyclopaedia*
4. *list read*

If we were to do a cartography of our prehistory, then the world would have had only two continents in the eyes of our earliest ancestors.

DNA Migration

The DNA Migration Map first introduced as part of the Genographic Project, launched in 2005 by the National Geography, has completely altered and moved history to a new frontier. Before I continue however I would like to add that I am not affiliated with this project. This project is publicly open to everyone who has access to a computer. As a researcher however it indirectly and unknowingly works hand in hand with my views on what is the incremental building of humans over time.

To the point when we look at the maps generated by this fantastic research it present the various patterns of movement; just one visual gives us clues that we may need to relook and rewrite every aspect pertaining to our global history. In our primitive past (as I say this with loving humour) we were always guided by the governance of leadership's account of history. Everything written therefore resembled a quote out of Africa, "until lions have their own victory the hunt will always be in favour of the hunter". But it is exactly these types of ideologies that human with progressive objectives should move away from. One visual of this project gives us the notion how integrated earth really is in design. In pushing human sustainability proposals which is our ultimate project our human connection needs to be highlighted in order to be a collective, for any push forward.

The notion that there should never be the need for anymore distinctions amongst humans, is no different than let's say Socrates enlightenment theories. Regarding western study of philosophy or any domain of critical thought, logic and rationalization. In addition it is the rationalization of why both books "Animal to Hominid" and "Let's Evolve" are so different yet interconnected and sequentially done.

When dealing with any part of history, every single type of analysis is by no means an easy task. Human advancement or simply put the dawning recognition that projects like this will aide higher education, is what I call a door opener for progress. Furthermore if we take this map generated by National Geography for example and make it slightly transparent then merge it with what I am about to present which is the study of linguistics or anthropology, something interesting comes about. Two different fields of academia both completely designed and interlinked out of our environments now sheds a different light regarding our earliest history. In fact our educational advancement moves up a notch since an illustration of our earliest ancestry will teach humans the understanding of what may have gone wrong with certain unfavourable past human social behaviours. This pertains to all aspects of life but most importantly leading up to the research on the social injustices done in the past and its connection to   primitive human requirements for environmental controls or even mismanagement.

As a researcher the biggest highlight that the Genographic Project provides us is how mixed these trails of DNA patterns are. The notion that we are one, starts our journey to progress. I call this the birth of *"real consciousness"*. Meaning if we pose the question what did DNA do that Law couldn't? Law is define as socially decided rules in keeping order. DNA will link us as a large family unit. A measure of our need to changing thought or morality *"on what is actually the value of life"*. Technology, DNA's study in aiding development, and advancements will no doubt come. Here is a quick preview on how DNA will change humankind, an inherent part of evolving, if we were to highlight our social responsibilities as those in an environment that are actually capable of doing so. I call this a form of measure for humanity:

Human Ethics and Social Responsibility

- Social responsibility - the idea that collectively, at all levels, what is beneficial to earth has to be decided and adhere too.

- Promoting technological advancements, DNA sampling in eradicating racism for good.

- What is Law at the global scale and are we really governed by Law, is this enough?

- Humanities role, in pushing advancements especially in areas of female empowerment.

- Ethics the governance to distinguish ourselves from animal. The critical role of unity.

- Assignment of value to any form of life, especially in humans

- Why multicultural countries are not enough in promoting tolerance. The push needed.

First Glance Visuals & Change

The divisional lines in our global map presented below will continuously be our guide. The focus will be our newest and our most isolated landmass, what I call Landmass 2 *"the Americas"* and our quest for dominant patterns in understanding our earth. As we work backwards, we will search to understand what really went on historically. Not DNA wise but rather the process of understanding how humans have transformed themselves in all aspects, with what their environments have given them. I call this the trails from the oldest to the newest topographies of earth.

Europe, Africa and Central Asia   **The America's**

Land Mass 1 (Intermix)   Land Mass 2 (Isolate Sampling)

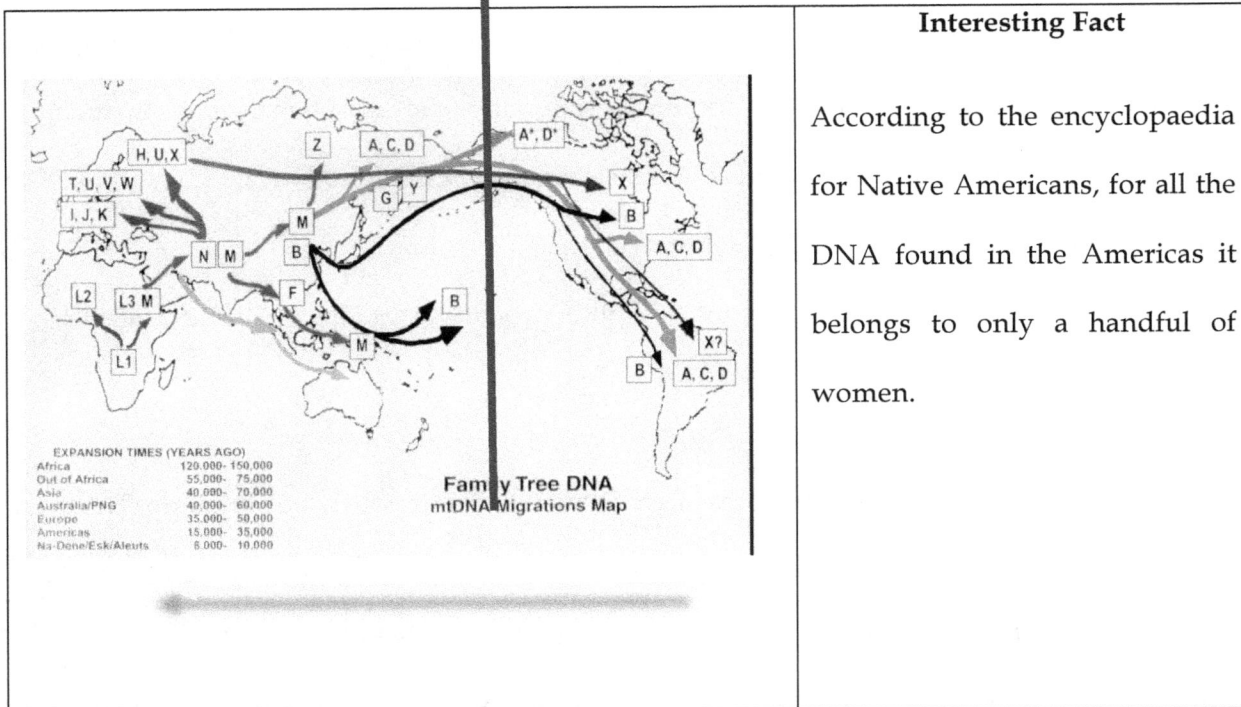

	Interesting Fact
	According to the encyclopaedia for Native Americans, for all the DNA found in the Americas it belongs to only a handful of women.

When we look at this map above and think of groupings of families that have crossed over does these funny lines all over the place make sense? Absolutely if you understand how clans operate. For each line in Landmass (1) represents a very close blood-line relation prior to crossing over. The interesting aspect is this is not indicative of the numbers that crossed over, it is just a sequential amount of chromosomes that are linked back to roughly around 9 Shamanic females that crossed, pre-population explosion.

The Revolution Behind Change

From the start of life to present, the table below is designed to illustrate the complexities of human integration to that of time. Awareness that of all of the topics listed below, each one are separate and is a perception breaker means the need for humans in understanding the differences in all . Let's go through them.

DNA versus Linguistics	⇨ Your DNA may be the same but your languages from birth may be different.
DNA and Culture	⇨ Your DNA may be the same but culturally or socially you may not be similar.
DNA and Religion	⇨ Your DNA may be the same but your faith may be completely different from other groups of people.
DNA and Region / Country	⇨ Your DNA may be the same but your region or country may be different to others in the same grouping.
DNA and Clans	⇨ Your DNA may be the same but your tribal clans may be different.
DNA and Physical Appearance	⇨ Your DNA may be the same but your skin coloration may be different.
DNA and Who You Can Be Related To	⇨ Your DNA may be the same, but DNA is so spread out like coloured marbles

	on a floor, that you may actually be related to someone very important.

Education and the Project Comparative Chart

The table provided below is a "Project Comparatives Chart" with the highlight of what technological advancements is doing and later on what it means for our stance on Human Development and Race Relations. Merging all three areas of research to illustrate why our world is changing is critical in understanding the evolving process to that of human progress. By analysing linguistics, mapping DNA, bringing in anthropology, understanding incremental building in the hominid to human transfer we can see how ordinary humans were merely a form of animal in their own perspective territories.

The chart below will demonstrate and show the connectivity and the importance of three different areas of research for advancement.

Project Comparative Chart

Project Title (1) National Geographic	Project Title (2) Animal to Hominid	Project Title (3) Endangered Languages
Research Format:	Research Format:	Research Format:
Mapping DNA	The study of the global movement of Linguistics out of Africa and the analysis of Hominid to Human incremental building. Done via observational methods in certain regions for resulting genetic mutations,	Preservation, of Dying languages through documentation and classification of linguistics.

	auditory sound adaptations from different topographies and attempting to categorize human visual profiling.	
Designed By:	Designed By:	Designed By:
Genographic Project 2005	Independent Researcher	Organization is done by UNESCO, part of United Nations agency.
Project Focus:	Project Focus:	Project Focus:
Is on DNA and gluing together the patterns of the movement of our earliest ancestors.	Is on First Clans and their earliest Shamanic Dialogues. How it all began, what was given to us from the animal kingdom as genetic traits. How language and human was designed from earth. Book Focus (2) Sustainable Management Proposals in highlighting social human behaviours towards change or development.	Supporting linguistic analysis to preserve our global language heritage and aid in many other areas of research.
Possible Issues:	Possible Issues:	Possible Issues:
DNA does not provide what early human adaptations, skin or human	The need for international collaboration and cross referencing to prevent backlash from academics. Due to the sensitive nature of what is skin color analysis and its historic	Media will eventually make uniform, our link to culture and linguistics.

profiling to topography are.	association to that of the study of first humans and their dialogues.	
Support for Change:	Support for Change:	Support for Change:
Social responsibility as humans in doing DNA for progress.	Email us a Shamanic related story of a region, or if a word is deemed as "Pictionary Descriptive", after completing this publication that word descriptive can also be sent. All of these emails will be pooled together, dissected and sent to the dying languages institute to be preserved. It also aids us with anthropological research. www.linguisitcheritage.com	Linguistic heritage comes to us from elders via songs and stories. Your grandmother is bilingual in a language for example she only understands email: address: Email dyinglanguage@heritage.com

The transfer of more rapid information and the research behind academic mega-projects like the ones presented above means an aid for advancement since technology enables us to be more enlightened as a species. I call this narrowing political difficulties to form a higher thought since our world is definitely changing as the basis of human morality keeps developing.

References

1. Prehistory and the first civilizations j.m.roberts v.1 1998

2. Genographic Project launched in 2005 by the National Geography

3. DNA analysis shows that Native American genealogy is one of the most unique in the world ANNA IEMIND

4. pink item

5. time magazine

6. Plato the last day of Socrates penguin classics

Chapter 5 - Anthropology Overview

Not knowingly, even Charles Darwin married his first cousin...Darwin Eh?

The study of Anthropology is important because it help us examine human beings[1] and the diverse ways that people live in their different environments. To understand the basis of this publication we have to go through a few areas that are critical in helping us understand the cause and effect of our own progress over time. Using Anthropology, the following few categories are a glance of the development of our earliest history.

Overview

> *Mutations "The Component of Life" Analogy*

> *Integration & The Man Who Walked Theory*

> *Collective Behaviours & The Birth of Culture*

> *Introduction to Historical Stagnant Populations Numbers*

> *Development of Language*

> *(Chapter 6) The Role of Topography - Pagnea and Landscape Comparatives*

For centuries religious scholars with evolutionist have battled fiercely over how we came into existence. This debate might have finally reached some type of medium, I mean Adam and Eve were one of the same human species type, weren't they? But would it change your perceptions if I told you they were related and very closely?

The suggestion I am about to present is in order to advance some type of median has to be reached between philosophies pertaining to God, theology and religion with that of the Academia of Sciences. One train of thought could be that God, we will call him the unknown, would have had to have started the process of life incrementally over time and scientific reasoning such as the theory on clan inter-mating would have had to have followed. The process of how humans came to develop into the numbers we have today starts with mutations. This process started way before any religious theories of enlightenment on earth even occurred. Approximately 3.85 billion years ago near the shores of Australia then attached to South Africa single-cell organisms started to self replicated themselves into all living things on earth today. A few single cell gave us everything as they exponentially grew. But it did not stop there because somehow another mutation followed. South Africa, if we study the most heavy trafficked or destroyed topography of earth became where prehistoric life actually began.

We are about to explore the study of mutations, how they came to be, their manipulations, how they mushroomed erratically in new environments and the transformation of these cells. It is exactly these mutations that can actually help us later on understand human social behaviours and how it has unknowingly hampered human progress. That the very nature of any brain regardless of its size or DNA makeup, is naturally designed to store any format of learnt information. It is exactly this back and forth transfer of knowledge, naturally picked up from its surroundings which enhanced the development of our earliest ancestors. Today what happened historically can help us in determining our own behaviours and how we can use it as a blueprint to recode as I say our very own thought processes.

Our goal eventually is to spark debates and understand what has happened in our past. The purpose of academics after all, is the forever ongoing process of growth . Our understanding of the minuscule steps taken from the animal kingdom to our earliest ancestral humans, is exactly the process needed to enter into the information age through the brain's transformation, building and growth.

This shift from animal mutations, learnt sounds to word conversions, adaptations to environments over time starts the beginning of socially driven innate behaviours. This is followed by development of full language, the start of relationships within social groups and our permanent mark on our environments. I call this the steps in the building of human.

To understand the anthropological portion of our environments role in designing human let have a look at these key concepts. We will used the following simplified analogies that will be referenced throughout this publication. I call this the next generation, a higher platform when medical sciences will cease in making species mutations or major irregularities fully stopped. The concept that evolving of human, within the species, itself will now germinate into becoming the new humainoid species.

Mutations "The Start of Life" Analogy:

Centuries ago a Chinese farmer noticed a slight mark in one of his school of fish. One single black fish had produced an obscure gold spot. The mark glimmering in the water, he wondered what would happen if he can produce an entire school of gold fish, to maybe give as a gift to the emperor. He immediately isolated it and let it keep mating. Then he notice from that single one black fish with a spot, another gold spot was reproduced. He now isolated both together and let them mate until gold fish were completely gold.

- Key Point - Like this very first spot of gold to appear on a black fish, our environments generated all living species, from a by-product of a mutations.

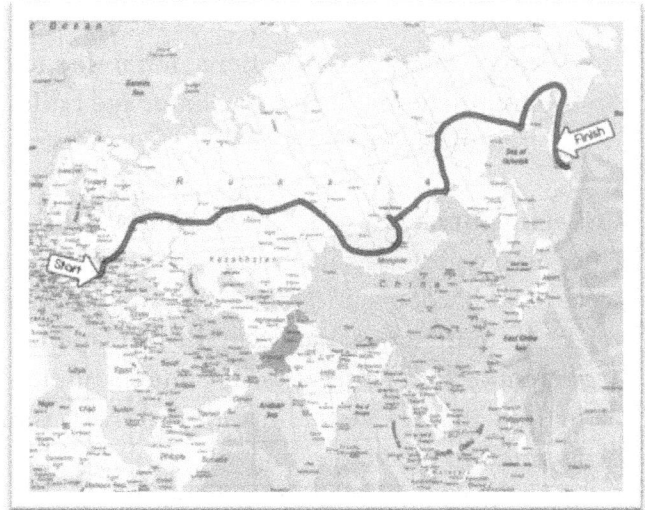

Integration & The Man Who Walked Theory:
The man who walked was developed into a theory based on a comparative of past and present. It was to illustrate how our earliest primate / hominid integrations, via mutations lead up to a species called "human" and how humans actually evolved. Using the nomadic behaviours of our earliest hunter gatherers in retrospect to that of a newspaper article about a man who (2013) lost his wallet and all his possessions in Europe and walked all the way back home to Kamchatka peninsula (north of China) is the debate. The irony is it took this particular man a little over two years to do it. This is based on a true story. Now envision our earliest humans thousands of years ago, following herds and waterways.

This theory can also work hand in hand with what I call species introduction in new landscapes. Meaning how all new varieties of living groups started. Our evolutionary journey, integration in different ecosystems, or migrations out of Africa as an initial crawler at the basic level of fungus and bacteria.

In series two, the immense pressure of politics over academia, our very own social animal behavioural traits carried forward in needing territory, creating visual and verbal distinctions in our earliest human species for territorial safety and why this basic common sense integration may have been hidden will also be illustrated.

- Key Point - When we look at the map above let's remember our earliest humans were merely a continuum of constant DNA variations and adaptations as they migrated to different areas?

   Awareness of the real component of time, to that of inter-mating clans who integrated means a major push forward not only regarding human progress, but for the movement of civil rights as well.

The biggest awareness therefore is that a continuous alternation had to have taken place in human and species DNA, over the course of thousands of years. This is especially important for the sections pertaining to brain development to that of adaptations, linguistic evolution, our language building to migratory movements and what exactly is "cultural identity".

Collective Behaviours & The Birth of Culture

How do we observe social behaviour in our earliest history? We had no written, all artefacts were biodegradable and there is no way of going back in time. One way might be to cross reference human behaviours to those in animal species. Remembering always the whole argument behind Animal to Hominid is human's should do a self acknowledgement that we still have traits that are very animal in nature. It is highlighting the development of our earliest social behaviours in pushing change.

From the start of life with rotational splitting behaviours in microscopic bacteria, the animal kingdom and our immediate surroundings have massively contributed to actually designing humans. This train of thought is not exclusive to the certain acts themselves but also in the transformations, birth of cultures and identities as well. This paragraph focuses on a brief overview of how certain social cultures were derived from our environments.

It is the understanding that once upon a time humans and animals were not a separate species and were delicately integrated in our environments. Take for example the behaviours or acts of huddling in species. When an elephant dies the remaining members of its herd immediately circle around the decease and mourn. In some cases using their tusks in feeling their relatives bones this behaviour is done for several days at a time. When a leopard attacks a herd of gazelles automatically they circle to protect their young. Using observations from different species in our ecosystem, our common denominators taken from animals which would have also been an innate part of primitive humans are the acts of circling.

These circles are not just a metaphoric shape. Circles represent self preservation in species. Attempting to bring back the dead, and / or the protection of young. These are genetically engrained innate behaviours that gradually, more like a few millions years later, converted to social culture for a variety of reasons. I call these two categories the circles of birth and death, which are what started our earliest cultural behaviours in humans. Here are some examples today of behaviours transfer over to humans from the animal kingdom:

a) Religious circular prayers in a chanting format today like Zikir is associated with Islam and is found in the regions around the first civilizations such as the North Africa, Indus Valley and Anatolia. The irony is Zikir predates Islam and is actually a circular style shamanic prayer format which probably started out with hominids grunting and mourning over a decease. How can we be so sure, for two reasons some Islamic clerks still consider Zikir as pagan. Paganism predates all the major religions. Secondly

58

watching a video of that format of prayer can bring you back 10,000 years if you simply

mute out the newly incorporated verses of Islam in a Zikir style format of praying.

b) Megalith structures such as "*Stonehenge*" (there are many), "*Gobekli Tepe*" , and other

circular columns which started off as human carcass holders to feed bird deities in

Aztecs times are designed from innate behaviours unknowingly picked up from the

animal kingdom. We are not referencing a few hundred years here, what we are

referencing is the missing component between the behaviours of animals, transferred

over to Shamanic humans.

What would be their attempts to build similar objects to that of their group's innate
behaviours. Behaviours of what they were immediately doing which was attempting
to understand death and circling for safety. Unknowingly our earliest hominid /
humans started building their social behaviours. The association of their primitive
humans thoughts with their first architectural structures?

c) Then there are the old Anatolian format of mourning the deceased, which was putting

the body on a flat stone table and wailing around it. Historically these tables looked

strikingly similar to Dolmens. The irony is Dolmens are not local to Anatolia but found

globally. Locations of Dolmen's are found in Ireland, Spain, India, Korea, France and

Germany. My guess many more existed but most probably got destroyed as pagan.

Dolmens are one element that will help us connect the dots that there had to have been

some sort of uniformity of our earliest humans and their cultural behaviours. Any

reference to our global Shamans and will now label as the era of 250,000 years of our

earliest human ancestry and their shamanic beliefs.

d) Ever wonder why each country has their own foods, customs and cultural dances?

How was the ideology of being culturally different born, even between neighbouring

regions? The answer is the individuality generated by social groups in a particular

environment that was isolated and the requirement for distinctions and self

preservation are the main reason social identity was born. This all falls under something known as social patenting. Social patenting therefore is defined as when the world had no media to influence an inherent trait called mimicking, which was naturally generated from our environment, global settlements created unique identity through generations of their off springs various developments.

e) Cultural folk dances, common in many circular formats groups such as the varying Halays of each region in Anatolia, the Fire dances of Celtics, or the Earth Dances of North American Natives are other examples. The understanding that our ancestors were not as complex or advanced in behaviour like we are today, helps us in understanding their earliest behaviours and forms of social entertainments. I call the end result of these social behaviours naturally transformed into variety of cultural attributes even down to their earliest forms of entertainment.

f) Various customary weddings can also be associated with the acts of circling mentioned above. Technically both are inter-related meaning protection of young became protection of union of couples. Somewhere the realization in humans that monogamy, straying away from all types of sexual promiscuity meant self preservation from certain diseases brought on certain customs and rituals. The understanding that humans were not monogamous in nature, transformed itself to a celebration of union, with gifts. A form of promoting monogamy. The birth of morality followed. We can therefore deduce that morality and enlightenment can also be built out of our environments from our incremental stages.

g) Feeding animals in designated areas, scraping off flesh from bones with finger nails then burying bones, under homes or moving them to other places. Cultural customs to North American Natives and those done historically in regions of antiquity are good starts in understanding the development of certain Shamanic behaviours.

o   Key Point - Lets remember the innate behaviours of circling transposed itself from the animal kingdom to humans, which brought on some of our cultural, religious and behavioural outcomes we have socially today. Circular dances (halays), architectural structures, prayers, human sacrifices, fires etc.

Introduction of Historical Stagnant Population Numbers

To understand the dynamics of what went on in the world of our earliest ancestors we have to look at a timeline of human population numbers. Historical population numbers can give us the dynamics to disassociate ourselves from what I call radical fanaticism or distinctions, by illustrating how integrated and related we really are as species. This is called understanding the *"Russian dolls analogy"* dolls that come out of each other or more simply put the consequential building leading up to the various human varieties found on earth.

In fact by graphically illustrating these low population numbers we can conclude that our earliest humans had to have been living in more recent times relatively balanced in topographical regions called not countries like we have today, but rather territories.

If we take for example the ecosystem in Serengeti park, Africa. All the wildlife in the vast kilometres that make up the park is balanced by mother nature over the course of millions of years. On a larger scale and in our natural environments our earliest humans were no different in the two Landmasses mentioned in the previous chapter, that they roamed in. With no real weapons humans acted as both predators and preys in the surroundings they lived in.

The Scientific American Journal indicates approximately a million years ago to 10,000 BC the population of earth was a steady number at 55,000, for our earliest ancestors. The hypothesis here being this number would have had to have naturally fluctuated but for the sake of argument we will always use the key figure of 55,000 hominids / humans based on what academia tells us today as the norm. For exact reference the period 1-2 million years prior to 10,000 BC, from a population point of view is what I called the "the *stagnant period of life*".

This was when nature had its checks and balances (pluses and minuses) in keeping primitive humans populations relatively steady, no different than that the analogy used regarding wildlife in Serengeti Park. In both publication this period is so critical regarding our developmental journey that we will continuously reference this time period. The diagram of "Exponential Population Growth" illustrates when the stagnant period of life started to peak but the focus should be in understanding how stable everything was prior to 10,000 BC.

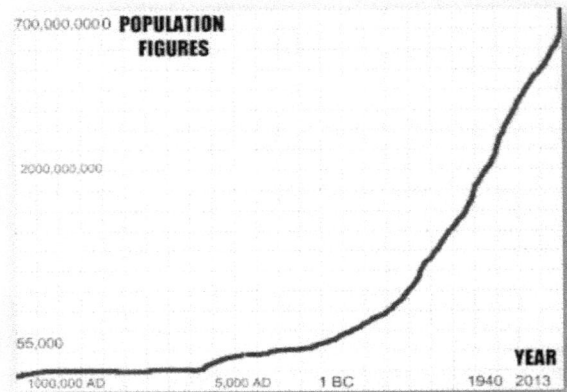

o   Key Point - The key in understanding human

incremental development and what prompted radically changes as a result of exponential growth

in populations, will later on aide us in understanding environmental sustainability.

Development of Language
Animals have communication, the problem is humans don't understand their communication. This does not mean they don't communicate or their language is not advance. It just means that we don't have the best capabilities or technology yet, of understanding what they are exactly saying. Attempting to chronologically put in order language and study descriptive linguistics opens a different type of window to our past. Lets never underestimate our earliest ancestors, this is inclusive of our primates or hominids. Language came to us from the animal kingdom and it is exactly these variations of sounds, vocals, that were transferred over to our earliest humans via their interactions with their surroundings. I call this transfer of sounds the base of all language foundations we have today.

The Brilliance of a Single Bacteria

Lynn Marguildis, Professor Geo sciences quoted the following "Until two billion years ck ago the earth's atmosphere contained almost no oxygen, and its supply of carbon dioxide had been nearly depleted by microorganism. This provoked a crisis for life, which is depended on hydrogen bearing compounds and carbon dioxide for its supply of fuel.

Then a variety of blue-green bacteria evolved a new trick: they used light to split water molecules into their constituent atoms of hydrogen and oxygen. The hydrogen was used and the oxygen discarded, creating an oxygen rich atmosphere. Eventually oxygen-consuming organism the ancestors of all the plants and animals alive today evolved."

Visualizing the single bacteria's capacity to develop let keep in mind the evolving format of reasoning and logic of that of animals in their settings. Historically their communication has always been viewed as extremely basic or practically non-existent because as humans our dominance automatically compares the animal kingdom to a way lower standard of existence on earth.

The key being that the act of communicating between species and human dominance for being on top of the food chain are two very different things. That communication at all levels, during all time periods that life has existed, would have had to have existed. And should never be dismissed scientifically as irrelevant.

☆ Key Point - Early dialogues and the building up of first words was an incremental process of transfer from the animal kingdom to humans. This is a direct result of what our environments gave us over the course of a few million years. When we look at the chapter called X Marks the Spot, it will aide us in understanding how language evolved from the first topographical visuals, descriptives and interactions of our earliest ancestors.

References

1- The man who walkd 2013 find article

World book a-1 scott fetzer company chicago 2012
2- Great ideas of science, evolution paul fleisher 2006
3- Scientific America news journal 2008
4- Science and medical discoveries evolution
5- check green
6 National Geography Pagnea

A large ant hill with all its intricacies, compared to the visuals from the sky of the city of New York is our example. If humans really think about it, there are no major differences. We merely are all a by-product of land.

Overview

> ➤ Pangaea & The Importance of Connective Topography

> ➤ Hominid Colonies

> ➤ Observations of Panama

> ➤ Ottoman Analogy of Landscape

> ➤ The Constant of Habitat

> ➤ Land and its indirect contributions

Pangaea & The Importance of Connective Topography

Pangaea is defined as the attachment and gradual separation of historical landmasses. Studying the connection of topography based off the map below of Pangaea's visuals will help us understand, the connection of what's going on from the actual living species use of land ideology. I call this a before and after look at the possible damages done by either our prehistoric animals, hominid colonies or human themselves.

For example: Take the word Sahara in Africa. Introduced in the later chapters, this is a Shamanic Pictionary descriptive word in referencing early humans looking for agricultural land. As an environmentalist we know that the Sahara was once a lush forest, which was deforested only 10,000 years ago. Pangaea's landmass cross referencing, when glued together aligns the Sahara perfectly to Brazil's region to that of the Amazonian jungle, (diagram three). The hypothesis here is we can now attempt to understand time to what large colonies of early hominids / humans in North Africa did when no petroleum products existed and controlled fires were the only format of habitat. Keeping in mind that controlled fires are 1.5 million years old.

Soil samples of the Amazonian rain forest will start us off in understanding the same type of impacts when top cover is lifted and exposed over an extended period of time. Its gradual change transformation, and the desertification that took place. Desertification has many theories of existence, and many will argue these vast plains of sands are beautiful and natural ecosystems. The problems is sand heat generation capacities, the slow change from lush, the actual diminishing of life are what I call the slow tumours or wounds of earth.

Furthermore deforestation is exactly what began the historical catalyst of many negative factors such as extremities of local fluctuating climatic changes of the desert itself, hardships on emigration or extinction of species, reductions in local fresh water supplies, birth of our darker skin indigenous variations to that of impact of slavery, its psychology and lastly a connection to all present day violence and poverty.

Consequently any globally found topographical assessments of deserts along with other landmass comparatives will also aide us in understanding the critical importance of all living species to that of surroundings, and their association to actually developing. Sometimes not always in the best of ways.

Pangaea

The study of land therefore is the prologue necessary to piece together what occurred. It will help in evidencing the following:

The Study of Desertification By using an analogy of deserts being dead table formations or the same as the wounds of earth, we are starting our journey by searching for where the most damages of land are. Examining the oldest deserts gives us many clues. These conversion from red like clay top soil transformation to sand or now damaged soils as I call it can be linked to the most activity by species or what is known as the heaviest traffic areas of the world. With the focal point being where mammal life started Africa Namibia's desert region.	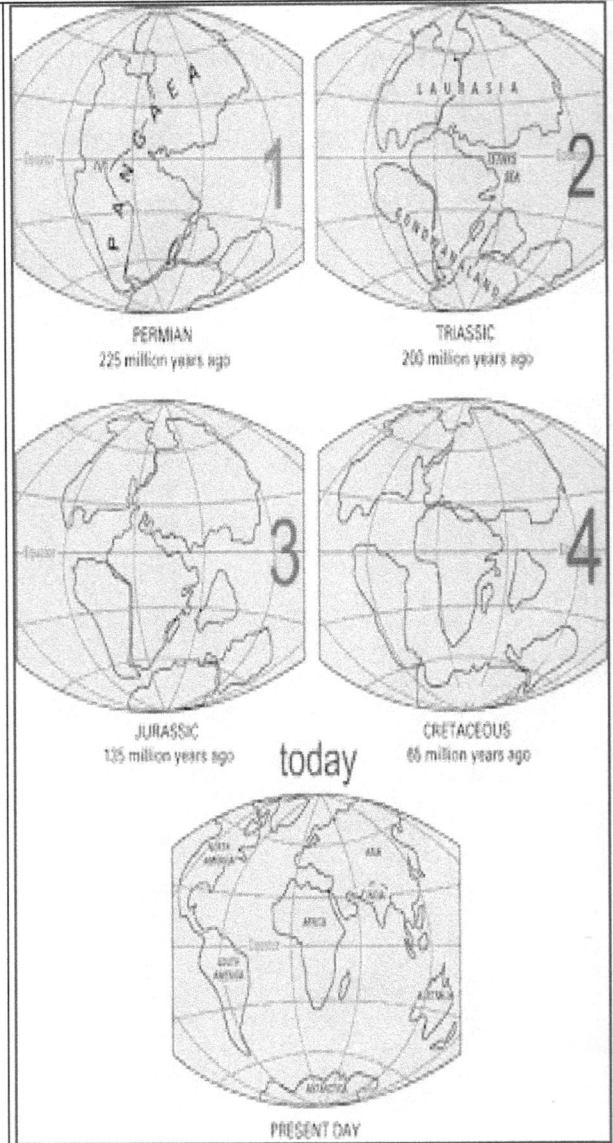 PERMIAN 225 million years ago TRIASSIC 200 million years ago JURASSIC 135 million years ago CRETACEOUS 65 million years ago today PRESENT DAY
Landscape Comparative Visuals Pangaea's land match attachments provides a before and after visual of the study of landscape based off age. Meaning in species to hominid to human	**Landmass Theory 3.85 Billion years ago** The concept that Australia (where bacteria was born) and South Africa had to have been attached prior to Antarctica's landmass lifting.

timeline what is new topography, to that which is old.	
Prehistoric Animals & Hominid Colonies	Monitoring Movement of Human and Linguistic development
For Environmental Conservation	Pangaea, Scaling & The Use of Anthropology

Removing longitude and latitude. It helps us scale earth differently. A critical element for many areas of academic research and divides up landmass (1,2) in order to see human development from a different historical angle. |
| Understanding Land | |

Indirect Contributions of Understanding Land
Invasiveness of Crawler Species
  ➢ **Location of First Civilizations**

Where the first civilizations were, based on destruction of soil and conversion to desertification in some regions North Africa being the start. Evidencing how fluctuating climatic conditions of Sahara and sand destroyed anthropological artefacts. No longer is Mesopotamia the highlight.

  ➢ **Land Contributions To Language**

How landscapes actually aided in the commencement, naturally altered or intentional changes in languages
  ➢ **Tracking Linguistic Evolution**

Which can also benefit humans in understanding linguistic evolution and word developments out of Africa. The study of DNA to Land to Regional Marks left behind.
  ➢ **Movement of Hunter Gathers**

We are looking at the connection between heavy use of land with aiding scientist track nomadic patterns of our earliest ancestors.

> **Social Impacts**

More so in series two, population grow hand its impacts;
Hominid Colonies

Observations of Panamanian Indigenous
As I travel through the dense tropical forested areas in Panama, I do a quick visual scan of a few central American indigenous groups of people, they are clustered approximately at around a 100 people per tribe. If we take that 100 figure divide by 55,000 (our critical stagnant number) as the steady number to work with, we are looking from anywhere between 450-550 nomadic tribes globally dispersed, between Landmass 1 and 2. For ease throughout the book we will use the median figure of 500 groups of communities to work with.
Our objective is to work with this figure, to give a visual of what went on in our earliest prehistory. Even if the numbers are not accurate since we can't confirm the historical population fluctuations that would have occurred, it a relative and good starting pointing of clans of families dispersed globally.

Land design

References

   1- The man who walkd 2013 find article

   World book a-1 scott fetzer company chicago 2012
   2- Great ideas of science, evolution paul fleisher 2006
   3- Scientific America news journal 2008
   4- Science and medical discoveries evolution
   5- check green
   6 National Geography Pangaea

Remember the game what came first the "Chicken or Egg", well finally we have an answer it was the Chicken, but a mutated kind!

Our surroundings are naturally composed of a variety of species. To understand the contributions of our environments to early humans we have to go step by step and comprehend what I call the continuous transfer of life. The transfer of life, is a paradigm and will provide answers on how we commenced? What is a mutant? What are clans and why do some communities still stay inter-married amongst themselves? How did they adapt to their environments? Why is the transfer of life important in terms of human development? These are key elements in understanding our earliest beginnings. Our goal is to look at the past to reshape and resolve current event regarding sustainability in our future.

Criteria's for research
The following areas are where the source of information for this chapter was predominantly derived from: *Reporting*   ➢ Southern Turkey regarding Clans   ➢ Central America Panama Natives   ➢ Nigerian clans, from the Nigerian community   ➢ Ulas Clan family analysis

Our First Female

When we look at Charles Darwin's famous evolutionary chart could we have taken it slightly out of context.

As much as we have a shared ancestry with primates there is still a vast gap in timeline between the very first primate and our earliest hominids, approximately 60 million years. As much as they are a development of one another in the end hominids became a new and improved mutant species. This is no different than attempting to chart out the evolutionary chart of a giraffe for example. A unique species that is left to earth with no real previous ancestry today. In the mind of someone who see intricately the true quantitative contributions of time, for fun we can then compare ourselves to a tadpoles and call it a day. The whole purpose of this book is to distance ourselves from actual monkeys, after all we are evolving are we not? As we attempt to illustrate what hominids (our true and closest family) have given us. Using different yet creative techniques.

Life started with a genetic mutation in the tiniest bacteria, over millions of years and many species later another alternation occurred this time as a type of bipedal primate. Most probably as a defensive strategy to see further by lifting upwards and to run quicker from predators. This type of continuous change in posture remained. Many uprights later a series of our earliest predecessor to any of the Hominids followed.

famous PIC 2 be find graphic artist

Academics indicate a little over 250,000 years ago somewhere in southern part of Africa another mutation brought to this world the first singular *female human and / or family*, no different than the mutation case study of a single spotted blackfish in the previous chapter. This female would have had to have passed on her human genes and created another human offspring who self-replicated within the group. Sounds relatively simple, correct? Wrong if we attempt to understand how complex human race relations and social behaviours have been historically.

For the purpose of simplicity, in understanding self-replication, let me pose another question. Can a painter reproduce the famous Mona Lisa painting identically at the same time, in its exact identical format, somewhere else in this world? The answer is no in the artistic world well this type of irregularity of an *"initial mutation"* just would not happen evolution wise either. That two identical species at any point could not have been made should not be just a hypothesis.

Mona Lisa is a good analogy in not only understanding human connectivity but population dynamics to the whole process of what is the art of evolution as well. This holds true with our association to the first mutation known today as homo sapient. To clarify when we are referencing the Mona Lisa analogy we are continuously referencing the single act of mutation itself and not the numbers of offspring that may have been born identically to a particular species that went off course.

I call this our "One Family Theory" the concept that only one human mutation was made and this chapter will evidence how the dynamics of life has linked our earliest humans two-ways:

> ➢ 1 Mutant - exponential growth of clans
>
> ➢ Saturation point, time and the extinction of species

This will also conclude there was born on this earth groups of species belonging to a single alteration initially. As they transformed themselves and grew just visualize what a growth-split is. Meaning growth in numbers, splitting of communities. Growth split and growth split again.

Interesting Fact - Still not convinced that's how we became 7 billion today, have a look at this interesting study which illustrates beautifully, exponential growth in humans. A Science Today article illustrates "that for all of the blue-eyed people on earth, they all came from one single female human. Meaning a single female, with a singular genetic mutation. (6) produced all the blue eyes of the world".

I am copying this statement per verbatim, and would like to add to this since I believe it is not really complete. Let's have a look at this chart which is interchangeable but illustrates beautifully the general concept of how blue eyes or on larger scale homo sapient with brown eyes came to be the numbers we have today. Possible outcomes of the following table and the order is completely irrelevant. Interestingly blue eyes, we are not referencing whiter skin, based on scientific research is a genetic trait that is only a little over 10,000 years old. The Art of Evolution is applicable to all living species today. After all we do have only one lion, one zebra, and one giraffe, don't we?

A single Female mutated	Produced several blue eye offspring multiple births. - immediate exponential growth
A single Female mutated	Produced only 1 singular offspring with blue eyes - somewhere in her line of offspring or in her own offspring another blue eye was produced who had continued the process. And goes with Gregror Mendel's Law- bringing back genetic traits up to seven generation back. Through inheritance blending. I call this secondary growth via multi-generational inbreeding.

Term Overlapping

Matryoshka dolls

All living things are what I call an overlap of the development of another species. The more simple comparative is that of these famous Matryoshka dolls. A *"mutation in a species"* therefore is just an alteration in a species due to close family relations and what our environments have naturally selected to live.

Whether it is walking upright, blue eyes, or other adaptational features such as protruded ears, augmented noses, changes in hand and feet alignment over thousands of years - all are a direct by-product of some type of alteration. The very first female/ or family if identical embryo was produced of all species born on this earth is no different.

The Start of Life Series

The following series of life was developed to summarize and help aid an understanding of human interconnectivity. The main source stem from research on Anatolian, Central American and African clans. In series two, we will continuously reference this process of coming to existence, especially in the domain of civil rights, human value/ethics, population impacts, human behaviours and our goals in environmental sustainability. What our ancestors did in creating distinctions for self preservation by not having the advancements we have today to survive and not really understanding the following:

One Family Theory

One mutation in a homo erectus created the first female homo sapient(s). During age of fertility this individual or all of these siblings if identical would have had to have reproduce and pass on their genes to their new offspring.

This is where it gets a little difficult to digest, but our objective is to solve the mystery of our singular lineage to a single brown eyed female going through mutation. It is documented that some women have reproduced relatively healthy kids from their own biological fathers or siblings (2). Scientifically it is also very common in the animal kingdom to have healthy offspring between animals made from a biological parent or sibling. Lets highlight that by having a child from a sibling or a biological parent may seem like something so completely repulsive in the human world today; but was common and widespread amongst our earliest proto-types. Mortifying yes, but isn't awareness education?

Inherent Social Behaviours

The truth is extremely older first civilization in populated regions still practice this format of same blood union since it is a natural *"inherent practice"* in their behaviours for survival. The concept of highlighting that inward social integration for the protection and preservation of a clan, or even species, has existed from the start of time. However through education and new blood law requirements in some countries, these clans have opted to intermarry with further same-line family members. This is an attempt to avoid any birth defects or irregularities. Some families have mentioned to me that they never give their siblings away to the fathers line of relatives but always to the mother side of the family. Overtime the mothers line has been a safer choice for pairing young couples together.

Then there are the other extremely old communities like the Anatolian Jews which within the last 60-70 years have in majority completely abolish this practice of same blood unions. This would be another example of humans through education changing social practices. Before communities that have a preference for intermarrying get upset lets remind them that all life began in this format. In the end the key is to highlights that "*as humans*" we are all related to a single female(s) and unlike popular belief the concept of human did not mushroom from different areas of the world. On this planet there is only one species called human and no other. We will discuss saturation point shortly, but If you can understand this basic theory, how we all began, our ancestral link to the African continent then you have climbed up a notch in human social progress.

Going back to our mutated female, this new group comprised of a mix of two species moulded itself beautifully within their original community. This would have had to produced several other humans whose blood-line being the same would have made them excessively still primitive in nature. This little section is extremely important in series two, when we look at the ideologies behind how the radicalization of language and culture occurred.

Understanding Clans, Their Birth & Dissections

Remember the story of the brilliant bacteria adapting. Over the course of thousands of years for groups of species to survive, they either mutated or dissected themselves from their original source when food became limited. This is no different for first humans starting their trek out of Africa. When they reached a figure of close to 100, they became officially a term I label clan, or more commonly known today as a tribe. How we derived at this number will be explained shortly. A clan therefore is a word that should be combine to groups of people known as "*first humans*". Being part of our natural environments, this is no different than a pride of lions, a troupe of Monkeys or a school of fish.

These were groups of early humans designed from our environments, who use numbers to protect themselves from extinction. Over time their growth, the vastness of the surrounding lands, searching for food, hiding from predators, getting lost and other predators marks the start of separation of our first initial family. The following diagram is an example of "Clan Birth, Growth and Dissection" in its simplest format:

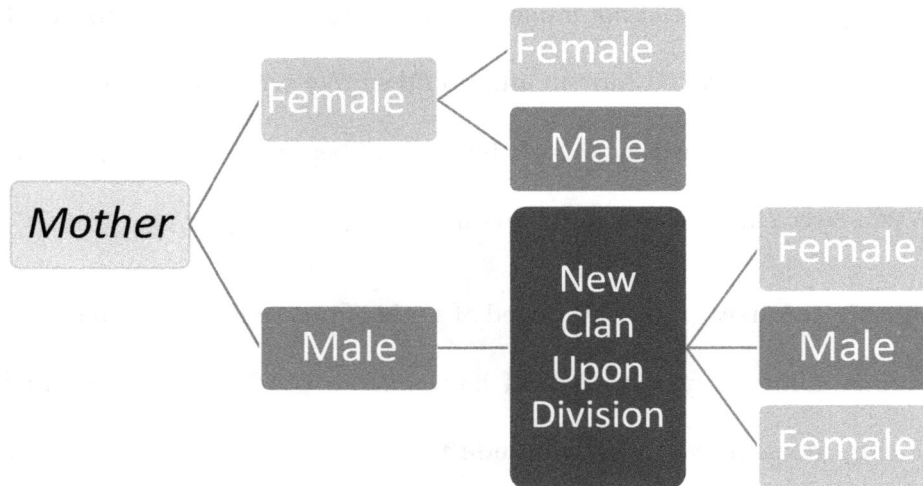

The gradual increase in numbers through inter-reproduction brought on secondary clans, and so forth. Over time all experience clan dissection. As they exponentially grew and separated, they started transforming their habitats into mini pockets of temporary settlements. Their nomadic migrations and the concept of natural selection during dissection allowed some regions to thrive favourable while others immediately encountered the harsh realities of a dangerous topographical environment.

Blending Academia and Theories of Faith

Religion for centuries dismissed the theory of evolution. In order to be progressive would real live case studies of groups of people intermarrying now change their vision of Adam and Eve? In the future would the theory of evolution be less of a taboo and more open to what research and

the science give us? The other objective of Animal to Hominid is to give a precedent to understand early human social behaviours and our environments. What prompted later outcomes even today.

This picture illustrates a Clan (Asirets) Gathering in Southern Turkey for example.

Interestingly their communal structure are very close in terms of mimicking social behaviours. Is this normal - absolutely, it is a combination of lack of social changes within these communities and their age.

This picture is a reflection of thousands of years of our early history but in this case they are wearing suits. We are fortunate these communities did not change, it gives us an understanding of the format of all our human ancestry.

The Extinction of Human Proto-types

As populations increased, and large vast lands of territories were now established by clans two occurrences had to have happened the first is our early humans took their dominant positions and killed the earliest predecessors to human and /or they mated with other Neanderthal, Homo Erectus or Denisovan types to reach a saturation point. Saturation point is the missing link that combines human dominance, clan warfare, time and inter-mating. Ironically it is not isolated to humans and applies to all species who are missing a traceable pattern backwards of their earliest ancestry.

☆ Interesting Fact - "In historical literature , it has been suggested that inbreeding was a major cause responsible for the extinction of the Spanish Habsburg Dynasty (1526-1700)". In other words even our earlier Royal families came from clan structures, or blue blood as some historically referenced it.

Understanding Saturation Point, in Hominids

Still not sure about hypothesis behind saturation point, imagine 2 or 3 soluble liquid color dyes in a glass, what is the outcome? They have naturally reached an equilibrium state no different than what happened to various families of first humans mixing in a much larger area. Imagine the glass below being our environments. Soluble chemicals for example have an immediate integration rate but for early human, this homogeneity or uniformity even down to standardization of chromosomes, took thousands of years.

Saturation Point - The Animal Kingdom Comparative

Let's look at another dominant leader the lions, who are famously known as the kings of the jungle as a good analogy. A lion's pride for example is composed of 2-3 males per territory. We are going to make the lion's pride comparable to earlier communities of human, using an example within our own environments. It is a way in understanding human innate behaviours and our animal kingdom connection. What do these mammals do when showing dominance and entering into a new territory? They kill other competing male lions and their offspring then mate with all the females. What do our valued academics say, they say "we have a shared ancestry to the animal kingdom" now let's try to evidence this behaviour in earliest hominids. Yes using lions.

Anthropological help using Language

The benefits of using language in later chapters will help anthropologist understand social behaviours in early humans. For example in Central Mexico we have a word that is regional for Pich, (where c is phonetically equivalent to ch in any of the Turkic languages), if translated it references the region where domination of fatherless children occurred. A common practice amongst non contact South American Indigenous tribes to kill any offspring followed by either killing or mating with the females. ref outube video watched. Could this regional word be coincidental possibly but further investigation into the world of first linguistics not only unravel our shaman history but investigation into other tribes illustrates the same type of behaviour.

In Africa the famous miniature head were not always adult heads that were cooked and made miniature but rather fractioning tribes whose ancestry was put on a stick in format of their children or younger teens. if we look at these two examples it gives us a clear picture that really our ancestors were no different to that of the Lion's world.

Interesting fact when we do word dissection and auditory analysis with what is labelled Birtur descriptives, followed by merging it with the study of anthropology we start seeing a better picture of what actually went prior to 10,000 years.

If we were to therefore define saturation point it has to do more with the equilibrium reached in earlier human species, family types. This would have occurred between the dominance of early humans positions in their environments, and their intermixing of various genre of earlier predecessor.

☆ Interesting Facts - The Ottoman Sultans never dealt with their own local women in the Harems. As a status quo to the numerous Beyliks (*larger settled clans*) that they administered they brought in women from other countries belonging to other communities of conquest. The conduct of the Ottoman Sultan's are examples of social behaviours going back thousands of years that were incorporated into settlements. Visualize their same behaviour but only in a more recent format.

The Transitional Period to Settlements

When several groups of first humans grew to over a thousand, this is what I call a transitional period. The "*transitional period*" is the period where some temporary settlements were able to counter attacks and stay in their locations they inhabited, while other clans kept moving regardless of size. The Celts, Hittites and Germiyan brother's clan (Germiyanogullari) are an example of clans whose forefathers belong to Europeans, all having once stepped foot on Anatolian soil. In Turkey for example they housed approximately anywhere from 10,000 to 50,000 people as a single predominantly blood-related community. During their attempts to settle, they did what other larger mammals do for survival they, killed and attack any other threat that came along the way. Lets understand that packaged meat only standardized globally in the last 70 years, before then the meal was in the hunt of the region that provided it.

☆ Interesting Facts - Today German historians still come to the Anatolian region and look for historical evidence of their earliest ancestry.

Adaptations to Landscape and First Settlements

For some of our earliest clans fresh water deltas by shorelines in and around the coast of Africa or the Mediterranean was the most comfortable and populated habitats. Unfortunately extreme violence, population increases, the desertification of Sahara followed by the politics associated to slave trade brought down the once glamorous African civilizations that existed. These settlements are the first civilizations of earth even before Mesopotamia. We as humans need to rewrite history to include this large land mass and its very vibrant history known as Africa. Let remember Sahara became Sahara because of existing civilizations and heavy traffic all around it. In fact Cartage, ethiopia kingdom of kush ....egpyt existed in shaman format way before Mesopotamia ever did. Puts pics.

The Stans

We are not done in our progression of Human oneness as I call it and *"how we have become who we are today"*. When these family of clans became several groups composing of a region and actually settled by adapting to a region, they now were what the historic Farsi people labelled as *"Stans"*. What does Stan mean, it means the standing place of a group of people in Farsi. In their quest to push Islam their caravans travelled back and forth on the silk road and they impacted regional communities, and settlements.

Stan conversion to Countries

In the last 200 years with our advancements for agriculture, and domination of lands these Stans have exploded population wise, and evolved into communities of people as we know it today. To help visualize lets have a look at the following names: Uzbekistan, Pakistan, Afghanistan, Hindistan (historical name of India) with many more scattered in and around Iran and Central Asia. I will be using the word Stan simply to evidence one parallel, of mergence of larger communities of tribes with no timeline. Eventually these types of larger settlements all converted to the various empires, dynasties and countries we know of today.

The Birth of Your National Flag

There are pockets of areas in the world that would be critical studies in understanding our earliest history. Key areas of more intense research by academics should be Southern Anatolia, East Africa, Indus Valley and the indigenous in the Central American region, just to name a few. The reason is very little change has occurred, even with the onsets of both World Wars these areas were immune to major changes or industrialization like the rest of globe. These regions are great comparative for us because the most pertinent information is the impacts leading to or after the World Wars. When the world radicalized with population explosion in terms of full ownership of land.

This series is a step by step incremental process of how National Flags and its sometimes associated ill behaviours were born.

From the start of life here is a list of the final outcome of where we are today.

➤ How Tribes developed to Stans then countries

➤ How our landscapes were carved out to be borders after WWI

➤ Fanatical behaviour development for countries with population increases

➤ The art of war, divisional tactics and the breakup of clan names to the birth of individual last names. Meaning in the last 100 years you were now forced to pick up your own last name, you are no longer Jane, John and Mary belonging to a clan name. You have now become Jane smith, John James and Mary Barnes.

I don't wish to touch on series two quite yet but would like to end off this chapter with one real visual of what it means to understand our actual roots as humans. It is also time to bring back up the Genographic project and the preservation of dying languages organization and how critical these are for human development. The goal of these two projects along with this research project is to push evolving to a whole new level.

The following picture is called Our Roots, and it is dedicated to Alex Haley's movie Roots .

| The contradictory analogy here, to that of Alex Hayley's slavery-based movie is, the illustration of the dispersements of the first family of clans coming out of Africa, understanding the component of time and our

"Actual Roots". | 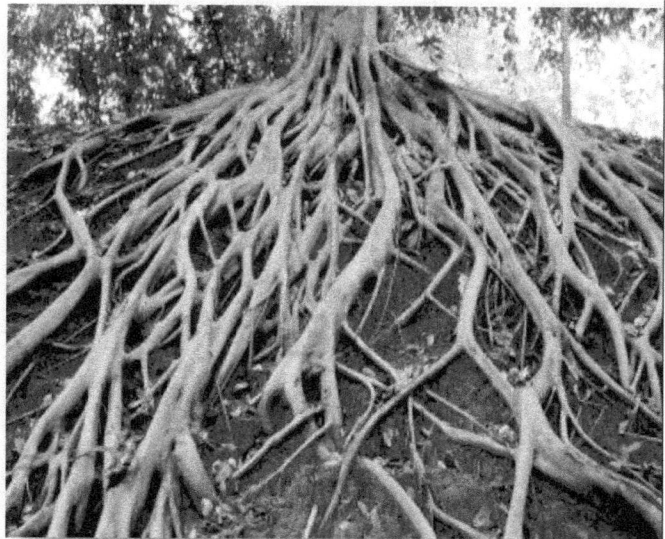 |

References

1. slave book find it

2. a book of negroes find it

3. anthropology book find it.

4. youtube video tribe kills other tribes children find it

5. mendel find it

Ever wonder why God made one of each for this earth? Well it depends on how you look at it, scientifically though the answer is the stabilization of predator-prey species over the course of thousands of years on incrementally developing landscapes.

Human / with variance is defined as one species only / called the thousands of years saturated one

Development : exclude social patenting, cultural behaviours- include constants of habitats for regions of antiquity say it professional = the different humans of earth today VOILA

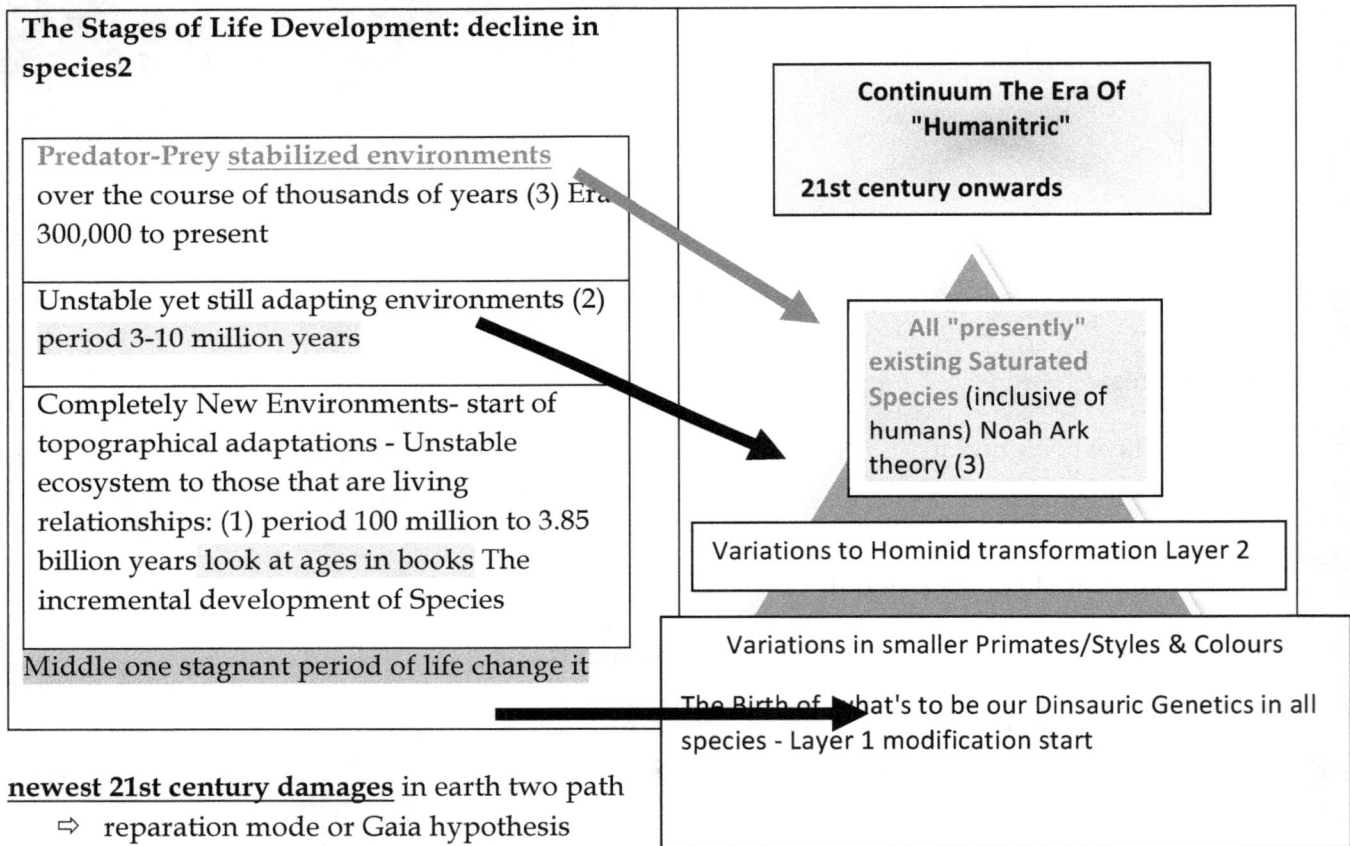

The Stages of Life Development: decline in species2

Predator-Prey stabilized environments over the course of thousands of years (3) Era 300,000 to present

Unstable yet still adapting environments (2) period 3-10 million years

Completely New Environments- start of topographical adaptations - Unstable ecosystem to those that are living relationships: (1) period 100 million to 3.85 billion years look at ages in books The incremental development of Species

Middle one stagnant period of life change it

Continuum The Era Of "Humanitric"

21st century onwards

All "presently" existing Saturated Species (inclusive of humans) Noah Ark theory (3)

Variations to Hominid transformation Layer 2

Variations in smaller Primates/Styles & Colours

The Birth of what's to be our Dinsauric Genetics in all species - Layer 1 modification start

newest 21st century damages in earth two path
   ⇨   reparation mode or Gaia hypothesis

   ⇨   destruction mode extinction

UNLIKE PREVIOUS SPECIES HISTORICALLY, NO GENETIC MUTATION HERE. BECAUSE OF OUR INVASIVE NATURE EVOLUTION GENERATING A 2 TIER IN THE SAME SPECIES-

FROM HUMAN WITH GROWTH OF BRAIN DEVELOPMENT AND INTELLIGENCE, WE ARE NOW CALLED HUMANOID OR HUMANITRICS IN 21ST CENTURY.

GENETIC OBSERVATION HYPOTHESIS good example: ulas family again as an example (same blood -genetic stimulation) **brings back dinsauric traits**
human tails analogy occurs indus and anatolia predominantly
(V)

References

1- The man who walkd 2013 find article

World book a-1 scott fetzer company chicago 2012
2- Great ideas of science, evolution paul fleisher 2006
3- Scientific America news journal 2008
4- Science and medical discoveries evolution
5- check green
6 National Geography Pagnea

Human DNA is like marbles scattered across earths floor, let trace.

The Evolution of Human Skin Color

This book title is very appropriate to this section on the evolution of human skin color, why because it is taking Charles Darwin's famous evolutionary chart and changing the colors on it. Which goes in hand perfectly with what he was trying to portray over a 100 years ago, of humans understanding, of what it means to be "continuously evolving".

From the start of time our social status was determined by our skin color. The sad truth is these ideologies have far-reaching historical roots stemming from the mode of production of workers in any agricultural fields to the impacts of centuries of slavery. In Animal to Hominid our goal is not to discuss the impacts of skin color or the overall economical ethics behind the psychology of skin coloration. But rather bring to light not only how your color was derived but how your physical features from first clans reinvented itself over the course of a few thousands of years. This is defined as human skin color and feature evolution, the adaptive changes humans have gone throughout history.

Criteria's for research for Chapters xxxx
The following areas are where the source of information for this chapter was predominantly derived from: *Reporting*   ➢ Geothermal Climatic maps   ➢ Standardization of color chart

> ➤ Genographic project highlight

> ➤ Regional top soil analysis, sand and Desertification - via Pagnea

> Limitations using local and technology only as sources to research

Furthermore in this section our other goal is to find a common denominator and see if we can actually categorize human physical features. We want to merge it together with the research on first clan, going back to our possible closest source of origin, the Bushmen in South Africa's. A concept easily traceable by simply doing a collage of human physical features on an actual map, known as the *"photo-trek from Africa to the America's"*. Which will be presented in the following chapter.

So how was the Bushmen over the course of thousands of years derived as a connector source to America's? As we put on an investigator hats the key is not just the similarities in physical features, that gives us a clue but also associating it with the patterns of the oldest desert in the world.

After all did every single humans not originate out of Africa? The key is where was the most impacts on soil? Ironically, my research originally started with book number two. This expanded dramatically when I realized a precedent understanding was needed of our earlier hominids accelerations, saturation and adaptations to human development today.

In my mind for humanity and race relations because of my multi-ethnic background, Animal to Hominid was really a one in a billion chance of every coming to light. What a blessing for humankind I say. This section changed from researching to cultural investigative mode. The following list along with the table immediately afterwards puts my entire spot-light right on South Africa. But how?

1. It started with the linguistic of the regional words in Africa, and a comparative to the dialogues in Native books. This will be discussed in more detail in later chapters but my entire research started off with an atlas and word recognition. Gluing together recognizable words with the regions all around South Africa's oldest desert.

2. Followed by countless hours of media streaming on YouTube of African tribes and trying to find physical and color tone similarities. I mean there is a link.

3. My environmental background kicked in searching for impacts of sand and desertification. The question being which deserts were the oldest in the world. Namibia came up as the answer.

As an environmentalist I know firsthand the routes most travelled and used heavily have the most damages along the way. To clarify what is being said let change another human perception by now looking at this table. I call it the oldest to the newest landscape for nomadic movements. Understanding this will give us the basis of creating a before and after visual of skin color analysis as well. To fully understand, the categories they have been broken down as follows.

Physical Features Chart

Started with 1 genetic human mutant (s)	**One uniform look via dominance of early humans-** no extreme climatic impacts yet so it kept a degree of uniformity as migration started. 250,000 years ago.
Nomadic Pattern and different topographies. This is known as different lands and start of regional territorial	This is the commencement of early **Human Feature Variances.**

ownerships / temporary settlements. It created the start of different human features.	
Oldest(Africa) to the newest landscapes(America) - The break in the Bering "The America's" with a handful of people have become an isolate and our case study throughout this publication. ⟶ **Purity of Native DNA and why I am linking it to an integrated ancestral Africa is our focus**	Using a simulated Native Feature as our source on hand, understanding Africa's early topography, passing the silk road, scrolling through hundreds of tribal pictures of variant clans including African tribes, we lastly pull out the Bushmen. **We start connecting the dots.**

The irony of the connect the dots game with our earliest ancestors out of Africa to the America's on human psyche is so intertwined with the concept of their color, just this statement of our connection to a black female may generate tons of offence or applause. Majority of humans do believe in Darwin's evolutionary chart don't they? The difference is I am highlighting it.

Add to the following statement time, movement of people and the following catalyst which generated human skin color. It will help be our precedent in understanding history, politics and socioeconomics in the following series.

Maps in total

1) First Clan Color Chart 250,000 years to 10,000 B.C use graphic artist

Human Color

Still not sure about First Clans being labelled as "collective colors" of a single race from browns to different shades of whites. Look at it this way Negroid is a race but technically there are over 1000 shades of light browns to black in Africa alone. Did you really think our Earth was this bland? What is being introduced is simply a new color spectrum or palette but one belonging to the first people of earth and the biggest indicator of this is not our damaged soil tracking, or the physical feature similarities but what language has given us. In chapter ...

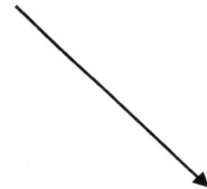

2) Today Climatic and Topographical impacted Color Chart (**Starting 10,000 BC to present**)

VULNERABILITIES

Skin colour map (indigenous people)
predicted from multiple environmental factors

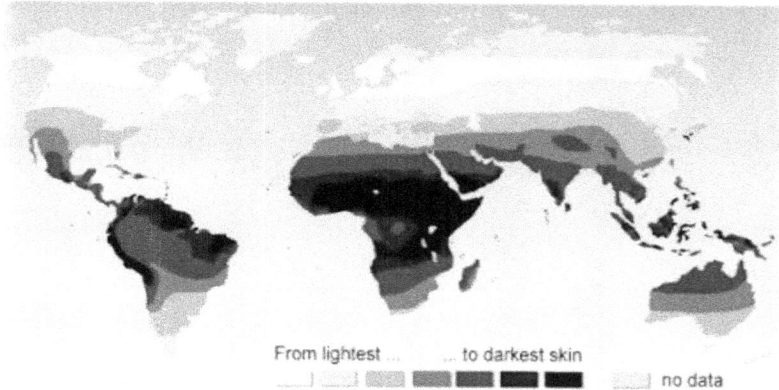

From lightest ...   ... to darkest skin

no data

ource: Chaplin G ©, *Geographic Distribution of Environmental Factors Influencing Human Skin oloration*, American Journal of Physical Anthropology 125:292–302, 2004: map updated in 2007

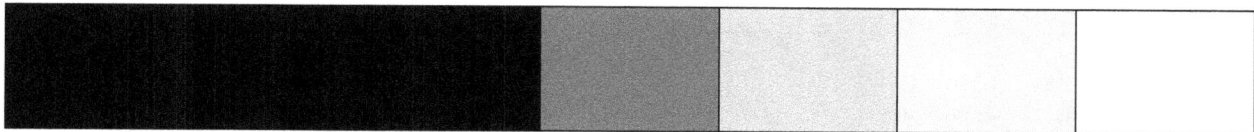

3)) **Find** geothermal heat **increases over time** map- **10,000 b.c goes hand in hand with the desertification of sahara. ck**

where is <u>the spike</u> put date my guess

4) **The future Color Chart- The America's if Amazonian desertification is unchecked.**

What is Human Skin Color

Human skin is the biggest living organ we have. It is an external trait and today we can classify it as follows. Your actual look or skin coloring at the present moment is based off the amalgamation of the following three boxes labelled 1, 2 and 3, with box number 4 being the difference in your DNA.

1	2	3	4
Understanding Time	What landscape gave and Topographical Adjustments	The climatic impact and changes of earth's Temperature. Rise in heat levels.	Saturation point -the mixing era, shooting out of new DNA from thousands of years. like popcorn

More simply put how was your actual color derived? Your skin color is based predominantly off what the sun has given you. Being so human right let my sarcasm at our history comes out, feel free to buy a "*I am Mr. Sunshine*" t-shirt for our global Landscape Betterment Charity and wear it with humour now. We are evolving. sexiest black model ebony wear it. yellow t-shirt yellow Ferrari background iam Mr.Sunshine.

As funny as I am attempting to make it if we look at our history humans went through the most horrendous abuses based simply off the long-term climatic impacts of temperature on their skin.

In the past if our academics didn't know about this research material and our single linear association to ancestral Africa, through clan inter-formation, of course the academics first analysis would be to assume humans came from separate sources. Followed by pointing out different physical features and trying to mould Darwin's chart with their link to the animal kingdom.

We can't blame the academics either, DNA wise there are also very clear differences and our scientist have always known this. DNA being what you are actually made of. Your very own genetic mutation started off with an irregularity that prompted recoding of genetics and / or an adaptive change.

When we think of color though lets understand the difference between box 3 and box 4 above with that of Genetics.

Some mutations that occurred were adaptive to topography while others were genetic. Let's look at both. Before we continue I would like to add "*adaptational*" is not a word found today in any dictionary but it is so well suited for the process of evolution (the concept of adapting) I have opted to keep it as a title.

Here is comprised the differences of what I have labelled a Genetic and what is an Adaptive mutations. The color of skin being an adaptive feature of course.

Adaptational Mutation	Genetic mutation
Topographical changes to physical features - Is a direct result of changes in landscapes when our earliest nomad migrated.	Comprise from same blood irregularities (the same genetic off combinations) - Related to Clans interbreeding. Applicable to animals as well.
Factors: changes in environments prompted this. Different foods, soils, temperature, intellegence etc.	Factors: Natural selection, saturation and only those allowed to live. Remember a daughter can produce a child from her father somewhat healthy.

Shorter time frame- our skin and physical features can change by 2 generations	Longer time frame- Millions of years. part of Natural selection.
Composed of externally related factors: ➢ The color of your skin, hair, height ➢ Desertification can impact eye shape ➢ What intermixing gives eye color variations like greens, hazels and greys. ➢ Humidity or cold the pointing or flattening of noses, lips size degrees. ➢ Social Behaviours	Composed of Internally related changes: ➢ Brown to Blue eye color mutation ➢ DNA ➢ Chromosomes ➢ Protruded ears and bump in noses. ➢ Blood types GENE SPECIALIST
Regular -Very Common traits regarding change	Irregular -Mona Lisa Theory (uncommon). Noah Arc story comes to mind.

☆ Interesting Facts - If we apply the rule of natural selection, understand human are one species and become aware DNA is globally variant. The question would then be, would smaller Central American pygmies be better adjusted to dangerous tropical rain forested conditions for example than their 6 foot, taller counter parts? The answer is yes. Height is also adaptational and based off nutrients and landscape.

Social Generational Patenting

All living species are a by-product of a mutant. But it didn't stop with one single art work of Mona Lisa. I would like to introduce what are adaptive social behaviours. This is more basically defined as the development or birth of culture.

When the human mind thinks of borders today immediately national flags come out. Especially if a country or its representatives does well, gold metals Olympics, innovations or one soccer match win. Are prime examples of behaviours associated to countries as we know of today. When we think of our earliest humans there were no borders. Ownership of land and its resources became a problem with the start of industrialization. Prior to then there were only territories no different than are analogy to a pride of lions roaming in Africa. This is also part of our Shaman history.

As we go on to bring out the world first language section, before major fiery or hours of entertainment becomes the explosive norm. Lets understand what Social generational patenting really means.

I will use the case study of Scandinavia as an example. If I were to go there and take 100 professors as my back up. With a very serious face tell them that they are now black natives with Turkic heritage what would be their response? Social patenting is a term used to understand the fusing of humans to their surroundings, in this case hundreds of years.

Groups of first clans not only adapting to a certain topography but also creating generations of social identity. This is a form of patenting of a social group. The Scandinavian would be offended or show extreme hostility no doubt because of their own rich heritage.

In all fairness to them the world of academia would be disrespecting their <u>offspring's hundreds of years of individualized, topographical identity.</u> When year after year change occurred, that which was different to anywhere else in the world. During periods when there was no real outside connection to other places, their ancestor left their marks. For Darwin he called it branching off from a primate. In more recent human timelines, if we understand the 60 million year ape timeline to hominid, I call it the branching off from a clan of thousands of years prior.

The internet is abolishing social identity and a global uniformity is coming out. This is debatable whether it is good or not. In the end globally our planet is changing again.

In the end Charles Darwin gave us his theories on natural selection and which species were allowed to live. Using his valued premise of incremental changes over time this research material gives us an overview of clans, genetic mutations, how skin color evolved in this planet and the birth of social identity.

In the layers of earth lies the bones of those who use to be here. The problem is we never once looked at the causes of hybrids and how its integration eventually became the development of the standardization of all species on earth. The Noah's Ark theory.

- The functionality of dinosauric genetics // Meet Kenny and Liu
- china gobi corridor and that africa

When we look at humans some academics might argue that the timeline between species will be an indicative factor of the existence of different proto-types of our earliest humans that had to have existed. Better defined as different shapes, heights, DNA and colours in hominids globally generated different humans.

The hypothesis here is there are several flaws in their theories,

- ⇨ the first - The concept of human was incrementally built -making hominids hence saturate amongst themselves and disappear. Understanding that we needed a foundation to become human. a human did not just arbitrarily show up in different parts of the world- linguistic time-lining is our backup in understanding mutations out of africa. what prompted language change needs to be studied.

- ⇨ Second is that the continuous requirement for the need of a natural state of equilibrium homogeneity in Landscape 1, of that of nomadic integration and habitat to human relationship is the concept of what actually not only standardized the first humans of earth but also gave way to different human physical features.

- ⇨ Eventually the two oldest landscapes (lush africa birth of hominid, and ideal resting point china water tributaries) became interchangeably in nomadic movements - remember population numbers to time. These became the first colonies that existed of certain hominids that started evolving completely isolated to that region via localized topographical adjustments and mutations. These two landscapes were the first hominid colonies

- ⇨ Mutations, eventually created another localized secondary saturation in our earliest humans. the kenny liu photo below. Technically because of secondary saturation all human ancestry should be traceable back to only two earliest human prototypes African Turkana hominid mix and central Asian pygmy mix- Australia being a British commonwealth country had a different variety possibly because of integration of slaves however there may have been a colony there as well and only dna will tell us. based off Australia's damages to landscape age

⇨ secondary saturation stands for when our earliest humans integrates with hominids in a

region and they pull in dinsauric hominid genetics from that particular of that region

example 100 newly adapted Asian pygmies 10 years later now in Africa generating

African pygmies or south American pygmies. historically confusing what is skin colour

to dna mix. the perception breaker no borders existed and population numbers were

very low- with a continuous need for nomadic movement in our earliest ancestors

This is what created the real source of variety in humans- followed by a third and fourth

etc. intermixing again. whites another mutation Caucasus regions

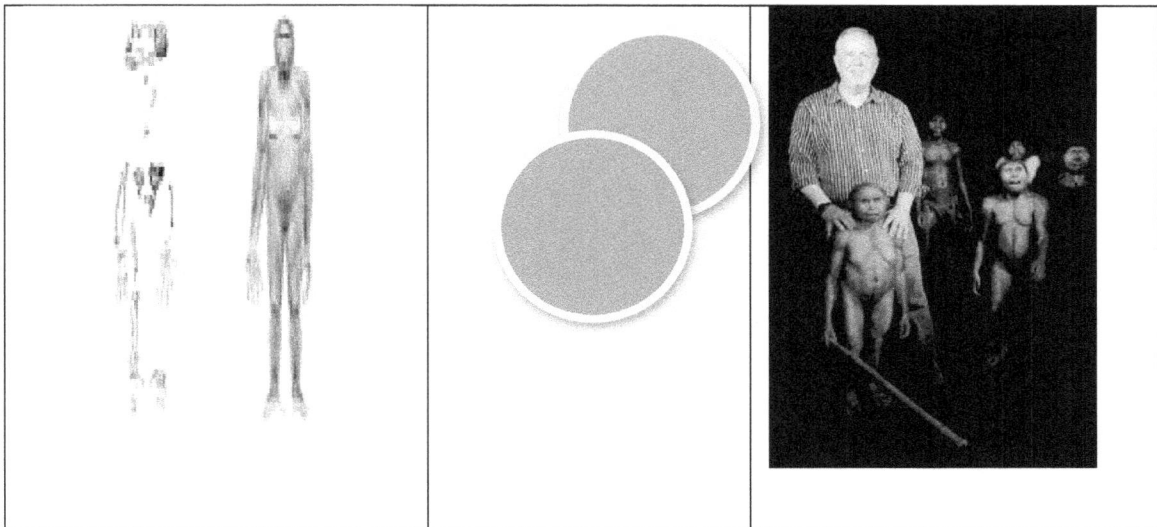

⇨ The confusion is multi-racial outcomes in humans and attempting to historically group

different humans possibly in colour related charts and physical features was not

properly coded. Color simply being climatic factors and not genetic, physical features

down to arm lengths being topographical adaptations. rabbit ear casestudy

⇨ For every 100 clan members of groups of 450 spread out globally (of our native case study group/ first clans) lets remember that would mean 450 genetic offshoots and mutations from each clan, standardized over time also creating variety

⇨ close interbreeding creates mutations visualize beak nose and a early hominid proto-type, or spots (today freckles) and another hominid proto-types // no different than a cheetah's spots -I call this what bones under the earth dont tell us. the clue is the different profile of humans today that tell us these things.

⇨ timeline means these invasive species out of africa kept producing different hominids all the way to china and then rotational backwards vice versa (genetic mutation within a group of clan theory)

⇨ integration is what made homind species disappear - humans came from one tadpole theory- we needed to have built human incrementally.

In this chapter we will analyze what it means for a species to die out because of mutations and integration.

is that extreme fluctuations in numbers of the distribution of species, mutations that occurred in each clan category, variety of intellectual development / use in the different topographical landscapes, and

This would have had to have been done via dialogue, mimicking and integration over an extended period of time. Or going back to one of each, Noah's Ark theory, this standardization is also applicable to all family of living species for that matter.

The concept of standardization is critical for understanding that Landmass 1 must have reached a balanced state of our earliest humans, pre-entry to Landmass 2 with an approximate date of 50,000 years. Why is this so important it shows America's connectivity in all aspects to Landmass 1 and more importantly to that of ancestral Africa. This connection being presented anthropological research throughout this publication.

Case Study

1 million years ago mammoth in landmass 2 disappearance is indicative of non adjustment to new habitats possible predator attacks

landmass 1 extinction via human activity

It also provides humans with a better understanding of how our earliest hominids disappeared. It is a powerful tool in human development, progress and promoting our oneness when it merged with DNA research or environmental sustainability initiatives.

The question is how can we analyse how this saturation was reached. Take for example the case study on hybrid grizzly / polar bear integration. Lets visual the future and how global warming has made polar bears reach the southern tundra of our northern landscapes. Eventually white polars bears will completely become extinct and the emergence of a varied grizzly offshoot will be born. When we analyze this to the historical disappearance of some species of earth we can then try to categorize what prompted a) the changes, mutations or offshoots and the creations of new life b) hypothesis or rational as to why may have occurred and c) what prompted the extinction in our earliest species.

The critical suggestion here is in each and every species born from that of the first tadpole proto-types out of water, everything on earth including items that may have been missed such as development of human linguistics was gradually built. Our development and very existence therefore is based off not only the disappearance of a species based on the strongest and fittest but rather the random building stages of a species. Meaning integration, nutritional intake in a particular environment, landscape contributions and that of eventual adaptation and controls of our surroundings are the contributing factors in making of a species to that of its habitat.

When we look at hominids we know their clans and integration are part of mutations leading up to early humans. Darwin's model therefore needs to be revised and made inclusive of integration and environmental contributions.

For every existing thing on this planet the nomadic movements and impacts on local species has been historically impacted by a standardization of that of its natural environments over time. The gradual adaptation or disappearance by means of amalgamation in new territory and intermixing of these new species in now older territory became another component of new species development or extinction? This is inclusive of the balancing in an environment and making of that of first humans. That environment keep in mind being Landmass 1.

If we try to analyze humans are they still in tadpole format? What are the key elements in a disappearance of any species? Hundreds of questions come to minds in the building steps of evolution including later on what are social behaviours or identity. Even in the 21st Century species redevelopment is occurring in our existing species but let's have a look at the list below and look at all possible reasoning for the stabilization or more presently destabilization of a species in an ecosystem. The table below is an attempt to rationalize species redevelopment.

Integration and Disappearance- The building of a species	Reason
Various animal species redevelopment (date)	Mutations - the birth and redesign of species based on new environments. New species in older territory and a mergence again based on their new environments.
Hominid / human clans inter- mating (250,000 years)	Mutations - the relative standardization of our earliest humans by form of integration.
Predator attacks (inclusive of some dissected groups of our earliest humans and hominids)	Elimination in species Extinction
Diseases namely bacterial infections or sexual	Elimination in species and the start of preventive behavioural changes (inclusive of human behavioural changes incorporation of religion)- Possible disappearance of some of our earliest primates.
Climatic changes to environmental topography	Mutations (less hair) or eliminations extinctions- emergence of woolly mammoths, disappearance of polar bears are some examples

Pollution	Elimination of species - acid rain and the impacts on aquatic species in different regions

The irony here is human social behaviours and Darwin ideologies incorporated for survival historically has always been the basis for slavery, colonization and imperialism. The distinction between human and animal separating and its gradual development with the disappearance of an earlier proto human type leading into the 21st century is the highlight therefore of what we will try to separate ourselves from in series two. Better defined as what made a species extinct therefore is also aid in understanding human social behaviours and our requirements for immediate change.

Leading up to Disappearance of a Species

Development of Clan Proto-types

We have incremental built and demonstrated the disappearance of a species through integration. This is over a difficult comprehension of the quantitative measure of time in the average human mind? The question becomes species overlapping.

When we look at hominids, there had to have been an integrated phase of transition. The concept of two of the same groups of species (one mutated and the other not mutated in a selective category) can co-exist until population swelling or further dissection occurred.

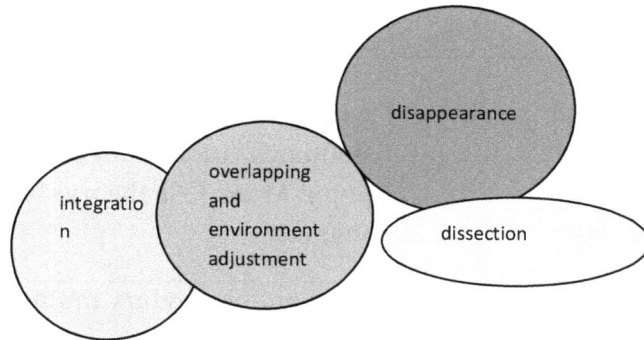

When we look at species overlapping, integration and death of a species why would these factors become so relevant regarding human development today? The answer is simple, the outcome of who you are today as a human is determined even in the same category of species by the virtue of developmental built social behaviours over time

- incorporation of mutation within a group of species. you look different but you are still us.

Social behaviours and the start of social acceptance and the start of variations within species of clans themselves (like the diagram above) leading to the birth of humans / mutations

> clans and the variety in them ulas family one walks on all fours they are accepting of each other

Gobi to South Africa Corridor- Lets read the land of our oldest colonies

first saturation human nomadic spread

secondary saturation human localized saturation intermix with topographically moulded hominid kenny and liu

historically confused in different humans with skin melanin differences

Level 1	Disappearance of various hominids / Charles Darwin theories strongest survive only
Level 2	Development of other proto-type hominid colonizes / localized adaptation to topographic conditions
Level 3	Human: now attempting to reach Saturation, with invasiveness behaviours and standardization point of a species with low pop / Mona Lisa theory- birth of our first shaman
Level 4	Time and Secondary mutations
Level 5	Incorporation of dinosauric hominid genetics in what is human and becoming localized saturation meet kenny and liu photos
Level 6	Hominid extinction / disappearance
Level 7	Human relative homogeneity still with hominid behaviours and localized physical appearances Shrinking of length of arms, height adjustments, intermix and variety creating dna differences over time (anatolian people is one dna) localized standardization for one specific colony region Jews african and palestian also have same dna Understanding time understanding dna

bibliography,

article bbc.

Temperature and understanding time plays a key role in human physical appearances. If this theory was untrue, then we would <u>naturally</u> have dark skin in cold countries as a norm.

include M chart

The funny thing about our vision of race is we are attempting to put 7 billion people in 4 categories of colors. White black yellow and red, that is if we search what use to be defined as race? In the past our academics did not have what we have today. Different groupings of people were still foreign to us and general descriptives were used. It is one thing to look over the hill and see a community of South American indigenous people in their habitats working, it is another thing to go to school with them and now actually integrate. Media, multi-culturalism in countries, organizations like the UN, ease of transport, a more fairness in socio economic balancing of different groups of people is changing the taboo of what was once foreign as other races.

The key to race has two underlying definitions of it. The first is what do you perceive yourself to be and the second is what can you be associated with in the event of an emergency. In the past this listing belonged to the group of 4 which was a very narrow category. Furthermore police would sometimes use their own mode of trying to identify.

In the previous chapter we look over adaptive behaviours in our earliest species. In various hominids that has existed, race was never an issue, It became a human associated problem very recently in human timelines. We live in concrete worlds, isolated so well from the animal kingdom that we have forgotten everything we have, including human skin color, came from a form of adaptation to topography. But let's break some perceptions, and bring out the real underlying factors behind race.

The Birth of Race

The study of climatic impacts on landscapes and the interactions of our earliest living species gave way to the following hypothesis. That our darkest indigenous along with our fairest Caucasians are both new to this planet and came approximately 10,000-20,000 years ago. The direct results of the impacts brought forth from the retreat of Europe's glaciers with the desertification made by humans of the Sahara created both new physical appearances. These landscape modifications are a direct result of climatic temperatures, changes known as the seesaw affect as ice melt and heat rises.

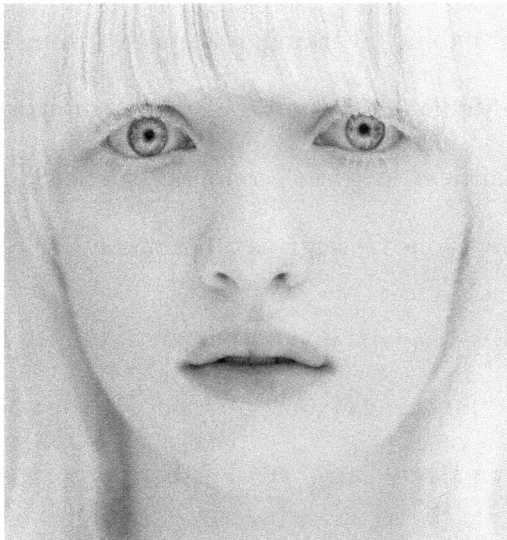

At the present the four general categories of races, can be broken down as follows Australoid, Caucasian, Negroid, and Mongoloid. And we mentioned earlier race is defined as a category of association to a set of people. Something you would tell police, during a need to be descriptive, without being politically incorrect. But with population increases and how we have adapted to our environments, can all seven billion humans really be put into four categories only. The answer is not anymore.

Adaptational evolving is a process that can occur quite rapidly not just for race but for physical features as well. Here is the list of seven that I feel would be the proper referencing to the seven categories of races of the planet today.

1. *Australoid, Australian Indigenous*

2. *Caucasian, White*

3. *Negroid, Black, African American, West-Indian or for Nova Scotia's black community, African Canadian.*

4. *Asian,*

5. *Depending on which side of the continent you are from Mediterranean or Spanish. * Large group of Hispanics, is the same as Negroid race, the lumping of groups of people when there are no other categories was our historical error.*

6. *Indian (referencing East India only),*

7. *Arabian (not Arab),*

8. *Descendent of First Clan or Native or Indigenous or any of the categories mentioned in previous chapters are all suitable. With the terms indigenous or native being the more popular choices.*

Majority of people should be familiar with numbers 1 through 6 on this list but the last one 7 is distinct. Technically when Mona Lisa occurred and our first human(s) was born, all humans should have fallen under the umbrella as First Clans of earth. However history made it convoluted.

There is a clear distinction between No. 7 and 1 through 6. A race becomes a race after many generations it has evolved to a certain topography not only color wise but culture wise. What is astonishing thought is that in some cases it is the speed of adjustments that is so surprising. We have taken for example rabbits in the United States and brought them to Austaralia and within one generation their ears have grown longer with their new topographies. Children of Mexican descent brought to the Americas are now reaching heights 4-5 inches taller than their forefathers have been for generations. Partly nutritional partly environmental both emphasize more recent change. The idea of race is included in these changes as illustrated.

Interesting Fact - In Canada the terminology African-Canadian is not really used, the reason is majority of the population of blacks come from the West Indies. These people a little over 300 years have created their own identities at all levels, with each island being unique. In Canada they are known as Jamaicans, Bajans, Trinidadians, Grenadians etc. Forget ethics and what was done, this illustrates how quick humans adapted to newer topography.

If I told a Bajan he is now African Canadian they won't apply it. Yeah we have a link but where is my flying fish and rum? Mi Bajan would be their response. Especially in festive times like the West Indian cultural parade in Canada known as Caribana when all the island flags scatter the city. Same with Jamaica, the island that supplied reggae to the world would never reference themselves as African-Canadian. Identity to them was clearly established via social patenting created by generations of island culture.

First Clans

Race historically was limited to color if we think about this moving a darker skin East Indian to Africa would not make him from the Negroid race. As a result the better definition would be, race should be a general descriptive to define a human to his inherited topographical region.

Consequently when we think of First Clans of earth this is actually a "collective race" meaning groups of people who are the closest descendents of the stagnant number of people that roamed earth, in a multitude of locations and slight color variations globally. Lets always reference their own color chart palette in the adaptational section. The realm is their history can be associated to any of the other races presented but is found in their purest format with the indigenous of the America's. Let's take it one step back and remember the component of time.

This is where we have to understand the differences between Landmass 1 and 2. Landmass 1 is actually an intermix of people over the course of thousands of years with the production of different DNA's over time. Natives and our South American indigenous on the other hand are astonishing, DNA wise since they are a regeneration of only 9 women's DNA.- they are the purest example of first humans. Genetic wise they were also very close bloodline related as well from the start.

Someone might say what about Mongols, Yakut and Bushmen people belonging to Landmass 1- my response is they are also members of First Clans, but would have definitely encountered some intermixing over the course of a few thousand years. Regardless they still fall under people who are descendents, people who are the closest to those of First Clans. This is equivalent today to a percentage of African Americans who were the earlier slave descendents and first transports from coastal regions.

Our focus in the next section now is South Africa's Bushmen and photo-trek of the closest to first descendents.

Let's have a look using historical race categorizations. Our case study is the Bushmen. Are we really going to call these two examples of Bushmen people a Negroid or a Mongoloid? If we were to focus on the existing.

© Survival International

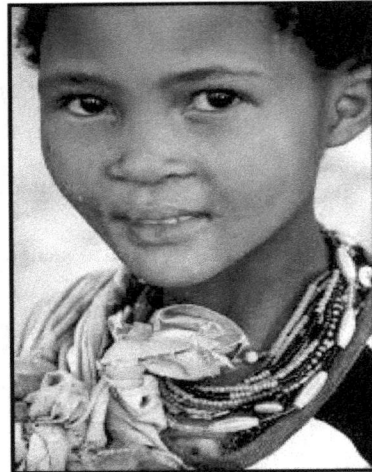

If we brought them to the United States, if they didn't speak at all and wore American outfits there will be confusion and they will be addressed a multi-racial. The answer is neither of the two races mentioned are really applicable. The uniqueness of their appearance, regardless if there is mixing with possibly the Negroid race makes the region and its people the closest we will get to our global ancestry. Hence why they should be known as First Clans Descendents or African Shaman Natives. African Shaman Native's history will be aided with the next chapter on the linguistic anthropology section and backs up these claims of this new identity.

Gaddafi is another example, he is African, he is a direct descendent of First Clans, and him staying in power so long occurred because of his large clan family. What color is he? Meaning during a police investigation how would you describe him if he was not Kaddafi. People passionate about African American history might scream and yell out he is a black man. But is he? Ok what about Yellow like Asian as they were historically known? Personally the concept of sticking 7 billion into 4 colors is humours beyond imaginable belief.

Answer: If he was just Mr. John Doe, majority of people would label him Mexican or Spanish. Now as the researcher I am cheating because I am using regional labels of south and central

America as my form of DNA and nomadic tracking, which helps as my examination aid. But in reality southern US or Mexican is really a normal description because he and his family are groups of people related to descendent of the first people of earth known as First Clans. This is a direct link to the people in this region that broke off thousands of years ago. They are coastal people, First Clans is a race of slightly lighter collective colors, belonging to the descendents of the first people of this planet. Understanding this was in periods when borders did not exist. Don't feel comfortable calling them First Clans call them descendent of indigenous. Both are the same. But Kaddafi along with Bushmen, Natives, South American indigenous are a very good example of people who are not Negroid and Mongoloid.

A little clue regarding Mongoloid's etymology by the way. They referenced it for people who looked like the Mongol empire or Asian descent, the category Mongoloid. Understanding this was right after WW2, during period of extreme hate. However it was meant to be derogatory because they also call down syndrome children Mongoloid. The difference here is I highlighted it for the purpose of proper etiquette and education. Asian or Indigenous even for Central Asian should be the accurate race descriptive today.

Race and Politics and Language are filled with innate behaviours of our violent history. My in-depth knowledge of cultures tells me one community that may be horribly upset with this research would be the Asian community. One is because of its slaves in South Africa and secondly some cultures want clear distinction with other communities which goes to cultural ideologies, by thousands of years. This is an understandable cultural norm.

My response is as follows yes Chinese descendent slaves were brought to South Africa but no the African Bushmen are not Chinese-Africans. They are still very indigenous compared to Asian descendent slaves who settled. In other words taking meat from a lion using sticks only, is not a settlement format of dining? It is thousands of years adaptational survival way before the economics of slavery even came into the picture.

In saying this let make something very clear to the planet, we need to acknowledge that all humans are of African ancestry. Progress starts knowing where we originated. My favourite quote out of Anatolia is as follows and is applicable to the concept of First Clans.

"The walnut came out of its shell and did not like its shell", goes to our paying respect to the connection to a single mutated first human 250,000 years ago out of Africa. That all races of this earth are associated with people of First Clan's. In the 21st Century it is time to finally put to its grave the discussion of race relations ever again.

Let add some humour to this sensitive topic that touches on social identity of groups of people.

White apes did not produce white people, nor did black apes produce black people. Yellow apes did not produce our Asian people. Lets visual this huge gap which then produced the first hominid. A Saturation occur with all our earliest hominids a gradual transformation to a uniform color produced (see the closeness in their palette chart) was our very first human. Then the First Clan color schematic occurred.

The only ape linked to humans therefore are called X-apes. If you want a great comparative ladies just remember all the X-apes in your lives, you will never ever forget this section, ever again. Especially in the area of teaching your children race relations.

Lets attempt to evidence now First Clan through photo-trekking. The process of taking certain selectively picked communities in both landmasses, and doing a photo spread throughout. Can you tell me where they are all from?

The objective always is using children before they grow. The reason being we want to evidence before they socially adjust and transform again to local settings. This is understanding "Genetics, to that of Adaptation". None of these children by the way are multi-racial. Our only clue.

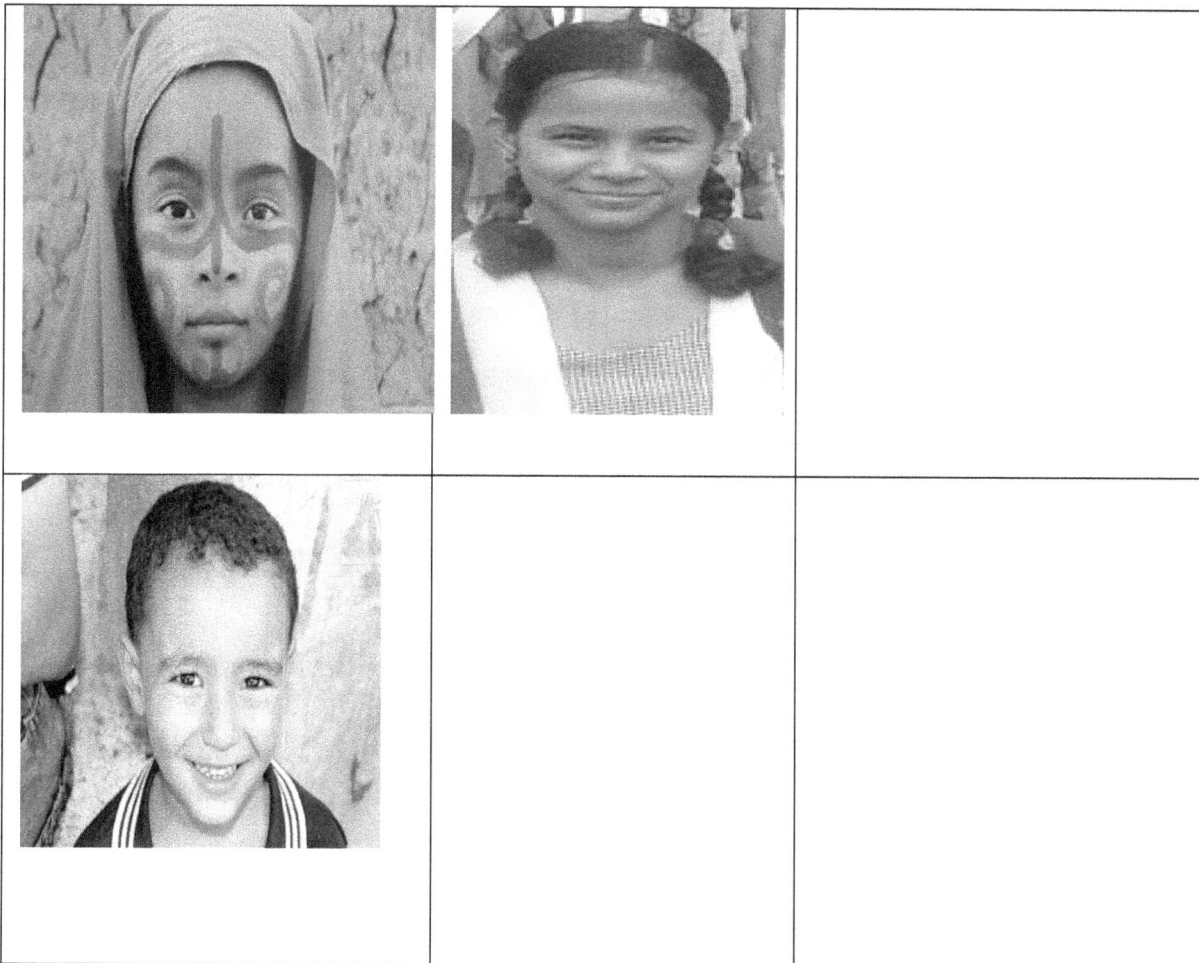

The answers are below as to where they all came from is as follows. Out of 12 how many got even a few right. In a test of 100 people the average score was......

If you are wondering how all of these children were selectively picked, several factors were taken into consideration 1) regions (where there is a lot of desertification), 2) coastal ways 3) linguistic analysis, 4) known first settlements of earth and 5) harshest topographies- where common sense would dictate that long term adjustments were needed.

A good example for point 5) would be the Siberian Natives. They also being in Landmass 1 may be the closest in terms of First Clan DNA. Why because of the harshness of their landscapes would have meant intermixing would have been to a minimum.

Regarding Siberia's indigenous, I don't have a laboratory and I am completely guessing of course. But here is why submitting in our DNA is so critical for advancement. It will push our interconnectivity and give us these answers. I am hoping this publication can promote the Genographic project as they attempt to pool together human DNA since this research in the near future will give us an insight of our connective patterns and movement of people. It will also help us destroy the social cancer we have called racism.

Our Darkest Skin Indigenous

The settlements in and around the Mediterranean, use of Sahara lush trees for fire wood retreated Europe's glaciers but this time reversed the roles of our first indigenous. Adaptational Mutations occurred to people with the desertification of the Sahara and the catalyst lead to the darkening of our first indigenous. Let's stop for a moment.

Doesn't all of this sound very funny now, the Greeks gave us the gift of enlightenment. I am using their theories and things I have learned in philosophy class and amusingly highlighting it for race relations. The awareness that topography and climatic impacted skin, would have never been a focused before. But it is desperately needed. When the first president of the United States became an African American, a southern slave state like Texas's did not even mention a headline for him in smaller editorial papers.

Caucasian Race ask doctor/ research

The start of the Aryan race commenced with its newer colder topographic settlements and Europe's massive glaciers retreating. A few hundred years later either from a direct descendent of a First Clans or within the Caucasian race itself an albino appear. Albino is a term labelled for a "genetic discoloration" of skin, its category does fall under Genetic Mutation as well. The first Albino of earth gave way to the repetitive genes of all albinos globally. Coloring balances back the gene permanently or temporarily in a human until another repeat occurs. 3,4 10 generations later.

The hypothesis here is that 1 mutated albino, in a topographically colder region created a permanent mark of all our blue eyed humans of earth. Not to be confused with all the other eye colors which are only a by-product of intermixing. Known as the product of dominant brown with a blue eye mix. The clue of number of blue eyes being in colder regions gives us the clear suggestions of where it was born. Traits are all dictated by dominance of a region. The interesting aspect here is that albino was also most probably born in Africa, and statistically more prevalent with the black community. Why the first mutant gene started with them and the age of first clan out of Africa (intermix). skin albino statistics ask

But the process did not stop there if the first blue eye was a by-product of an albino. For the offspring of female(s) with blue eyes to produce a color generated blue eye that would have meant an intermix and readapting again landscapes wise. It would have had to reproduce on a continuous basis an adapted, physical feature to colder climatic conditions. This is an equilibrium that took place rather rapidly in human time line which is over the course of approximately 10,000 years. As indicated by the relatively newness of blue eyes by our academics.

Melania Trump, V. Putin and the late Joan rivers, are a mix First Clan descendents, their skin may confuse us but their eyes give us a clear indicator. But let's watch their children, depending on their region or spouses their First Clan heritage will then readapt itself accordingly. This is no different than what happened in Landmass 1 a few thousand years ago.

Asian, Indian and Arabian

Skin evolution overtime of our earliest indigenous and modification to landscapes created newer race categories. Indian, Asian and Arabian are the end results. From an archaeological perspective these regions for us are extremely old, from a hominid-human time line perspective all three races are relatively new in terms of adaptation from First Clan to their existing territory.

Skin Color Outcomes- A General Overview

First Clan to Asian- influenced by Climatic and Cultural Impacts.

They are known culturally to stay away from the sun, under shade for hundreds of years. The painting of the face to white is also common in folkloric dancers.

Another good example to illustrate this, a Canadian-Asian descendent friend of mine went to an island destination to look nice and tanned before her trip to China. Her family was disgusted she was tanned. Being Canadian she told me about the cultural shock she went through. The confusion with skin color and ethics will be illustrated in book 2.

First Clan to Indian -influenced by Climatic and Intermixing. Slave import and climatic influences generated the numerous colors in India today. There is also a theory that blue eyes were brought to the Kalash people by Alexander the Great. Making them Phenotypes darker skin with blue eyes. Years of integration of empires.

In some cases lighter skin was a direct result of the British intermixing with the locals.

Pakistan is even a better source for First Clan as well, religion may have destroyed a lot of shaman history as pagan however being part of Indus valley it is a hub for Shaman research.

First Clan to Arabian - influenced predominantly by Climatic and topography.

What we are attempting to do is not discuss culture, ethics or oppression that integrated or changed people. We are simply illustrating the various colors of this planet and some scenarios of how they were born out of First Clans. These are only a few, in reality there could be a million reasons that lead to the balancing of these people to their regions, over the course of thousands of years. This saturation point of regional intermixing is why they are known as the races they are.

Australia's Aborigines

This region is a detached large island but why then are the physical characteristic so original for Australia's aborigines compared to other indigenous. Their look is very unique globally. Understanding when we are looking at race many factors are analyzed, including more unusual items like language.

Let go through their integration. Australia is still part of landmass 1 - integration would have had to have occurred tribes coming over but most probably the hypothesis is as follows since

Australia's a commonwealth country it is associated with slave trade. Slaves may have been brought in and other forms of mixing occurred with locals. This is called the transport of slaves intermixing.

Australia experience major desertification and since it is an island physical features overtime changed accordingly to that topography. A mix of slave varieties, now desertification and being an island transformed them to the unique characteristics they hold today.

The only reminiscence of their shaman history that is left is their mountain Ulu (Great Power). Ulu is a classic proto Turkic word and the fact that new Zealand and the rest of the neighbouring islands are agglutinative like proto-Turkic, also evidences that Landmass 1 along with south pacific indigenous were once related.

A thousand years ago our ancient people said "if you came from the skies be human".

our hominid dinsauric

So what is human? It seems the world is so black and white we forget everything else that exists in between. In the previous chapters we learnt that what separates people, their culture and social patenting is understanding time and environmental impacts. Hominid's transformation to first human took thousands of years. Followed by a relatively stable, period of existence for First Clans. Our objective is understanding this period and evidencing it to the planet using alternative methods. Here is the "Order of Life" it focuses on the breakdown and gives us a start on trying to answer what is human.

- ➤ *Animal to Hominid / Brain Start*
- ➤ *Hominid to Human / Formation of Intellect*
- ➤ *Human to Human / Developing period of Intellect*
- ➤ *21st Century / Information Age*
- ➤ *Comparative Ethics*

Over generations academia's biggest continuous confusion, was due to their attempts to merge animal and human. This brought forth the complete lack of thought to what was in between these two species or the period of time that had elapsed. The concept of intellect of hominid was absent. We know from previous chapters everything was incrementally done. Evolution therefore is applicable to all subjected walks of life in an incremental process.

So much that our requirements to understand these steps towards the definition of being human effectively broke down protocols in our own history, even during the height of industrialization. Especially in area as, race relations, win-win socio economics and environmental sustainability.

Definition

⇨ Human is defined as a single source to all species out of ancestral Africa. Period. It is in the event if you need a rare blood type, the understanding that it may be derive from any four corners of the world. It is a concept of understanding. The reaching of balance of all skin colors per region and a saturation point of all earlier species before us.

⇨ Being human on the other hand is the idea of putting a value to a human life. In series two it is the awareness that one born healthy, is better than 10 born to a ailing environment.

The confusion Period, Not the Confucius

Over time human have on a very serious playing level attempted to make the humour of the previous chapter stick. By creating all kinds of bizarre ideologies of what is defined as first human. The implication that whiter apes were the Neanderthal, Denisovan were the yellow apes, and everything else out of Africa was a form of King Kong at its best. The area of unknown did not stop there. Hence our confusion tacked on the confusion of race. You are some type of Mongoloid, I am Caucasian I am even more confused. Plus our DNA is different.

The start of brain is a look at the building up of what is human, to the present moment. In the end our earliest species did not experience race relations or exploitation of any sort. They ate, slept and mated. They only attacked if they were threatened or needed food.

Our goal is to lay a precedent in understanding human behaviour. **We need to understand the concept of what human really is?** This is critical for us especially in series two. Fights in Sahara for wood, to fights in World Wars for resources, Ottoman break up, Operation Torch, complete environmental lack of awareness up to the present moment. Especially if we want to look at the politics of Iraq, Iran and the North Pole more recently.

The highlight of something so simple as lack of experience in field of overall environmental management on behalf of our political administrators; is traceable back to the start of human population explosions. An analysis of the animal kingdom begins this process and shows us the stages of where we can head.

Animal to Hominid - Brain Start

Brain start is defined as the concept of the brain starting creative ways of adding on functionality roles in our earliest species. From the basic, an otter cracking food between two rocks or the mimicking of noises between animals it all began with adaptation. To the more complex levels, a chameleon adapting to colors in his environment for survival. Yes extreme stress did occur in our earliest species to generate something so astonishing like even color change.

Brain start was a form of building. It is the generational transfer of information from our animal kingdom. Certain interactions from the animal kingdom was so profound that we still carry some innate behaviours going back to our animal world. This is why awareness goes hand in hand with progress.

Hominid to Human - Formation of Intellect

The first real formation of intellect did not come from the human world it came from animal kingdom. One documentary video of monkey's socially hunting down a single chimpanzee will gives us the perception that intellect is way older than we think. Social organization is the base for intellect, it means premeditated dialogue had to have occurred. But the real formation of intellect came from our earliest hominids and here is why. Their start of controlled fires and the use of carving stones weapons for actual early hunting gear. Should evidence this accordingly. But other events in human timeline should also evidence this. Approximately 120,000 years ago in Spain cave painting were found with what looks like head gear evidencing another aspect of what I call our missing shaman history. The process of social gathering and decor. When external objects of benefits are used, painting are left behind, there was very organized thought. The last component of this research analyses is we know they did not have written but could there be something else they left behind that was language related? The answer is yes in the following chapter.

Human to Human - The Developing Period of Intellect ck dates of dolman / sanskrit

Around the time of 10,000 BC before the start of the main religions something mortifying happened. The origin of "I think therefore I am was born". Human intellect did a miraculous split conversion to higher consciousness, and our first written was born. Followed closely by other series of event, improved homeopathic, reasoning, rationalizing, philosophy and enlightenment. It is exactly this period that people started making the transformation from shamanism and started taking up permanent settlements, a far transition from our nomadic territories of the past, with a few leaders. Intricate detail was born. Meaning the world of arts, decor, literature, crafts etc. Unknown to the world Africa's coastal regions were now littered with their glamorous civilizations of life thanks to the natural resources all obtained by Sahara's lush forests.

Our intellectual capacity in innovation, to cure ourselves and the need for resources for comfort created population growth. The largest ever settlements was known later on as our *own Stans*. With the start of population growth especially after industrialization era, domination, ego, self preservation, and the need to use human for manual power began one of our dirtiest period in history. What I call a reversing back of consciousness and the start of a very advanced animal to animal combat. A vision then of what is progress.

This is inclusive of the modification of human history by our forefathers. A mistake I believe the world of our valued academia has desperately tried to fix but was faced with a dead stop, from political interference. Today part of this interference goes back to the problematic middle east and its black gold. No different than one era before the Ottoman Empire when our forefathers didn't know the extent of damages that exposing dead top soil of the Sahara would bring.

During its biodegradable shaman years this region was haven to people who did have language, culture and the arts. How do I know the clue is in the next chapter or for other more recent examples Native Culture. If you are wondering why I labelled it animal combat in the section on human to human, this is to bring out that we have not fully matured yet as a species.

INTEGRATION historical prespective only to generate new types of human

DIRECT -INDIAN CHINESE adjoining communities integrate tatar turkic and russian or from overlay

INDIRECT - CHINESE JAMAICAN CHINESE CARRIBEAN SOCIETY aboriginals australia/2 topographically adjusted communities are merged together to form a change

multicultarilism is different

The 21st century

The very start of the 21st century should be known as the transitional stage of our road to progress. The period between advanced innovation and not understanding the implementation for change to the confusion of innate behaviours in humans. The remarkable thing about this for humans, is that innovation brought forth the rapid transfer of information.

The following lists were all derived by the vision of our forefathers on what progress should be. But there was a heavy toll humans paid for this. Whether it was women, the impacts on our environments or race divisions we paid dearly.

Information age, genetic research and medicine, technology, transport, use of machinery, full access to scholarly books, discussion forums and aide, the transfer rate of information, cultural videos, photographic or visual accessibility, and the *"7 billion solution click syndrome"*.

Luckily for us today the biggest difference between industrialization and that what the information age should bring is ethics, use of technology and the breakdown or really understanding race.

Comparative Ethics

During this transitional period humans have to understand that the world is changing. The biggest challenges we face are not our own lack of innovations but the innate behaviours from the animal kingdom to humans that has not kept up with this new changing world.

The question then is what will aide progress or help us distinguish ourselves from our forefathers before. Believe it or not, possibly one never thought of before scenario. I call it the plus for humanity. The mixing of race and technology. Defined as breaking of taboo or what it can bring to other countries that don't hold multicultural regions.

Your polish son is now married to an Ethiopian in the US. Is one example. Lets add internet-conferencing to this, so you can talk to your new grandchildren. It is defined as being more tolerant and understanding race. Understanding our connectivity as humans as we push for progress and understand we have major environmental challenges ahead.

Being sensitive to humanity is also a great starter. The development of global ethic and civility is another. Actually separating ourselves from the animal kingdom, something our forefathers tried to illustrate 1000 years ago. Using technology to our benefit, no more slaves needed thank you we have our machines and ethics now. Did I also mention leaving countries underdeveloped is also a thing of the past?

The new Multi-Racial Race or is it?

Countries that have multi-cultural people will teach the planet what is human. DNA wise other than our indigenous in the America's we technically are all derivatives of a multi-racial ancestry. Let's carefully dissect. Whites ancestry is formerly out of Africa is it now? Bushmen being in

Landmass 1 had to integrated locally in the 250,000 years they have lived there. Anatolians integrated with different empires continuously, including the Mongols. Which would lead to the question what is Multi-Racial? As a researcher I am attempting to make First Clans as a race, can multi-racial not be in the category of race itself? The answer is no, and here is the justification as to why?

Multi-racial is defined as a temporary change in human skin coloration due to intermixing. It is a state of color in which over a short period of time it will evolve again. As a review what is race? It is the topographical changes in landscape and impacts of climate, that generated color in humans over time. It is a long term very stable process and its association to a set of people.

My children are of multi-ethnic heritage. For an outsider just looking in that would mean the first visual of a couple's children, of black and white union. This is inaccurate, the last count generated 11 countries my children can be affiliated to. They asked me mom "what race am I" one day. My response was simple whatever you feel comfortable as. I meant this not to cause confusion but full sincerity.

Bibliography

1) article slavery

2) tribe youtube video countless hours

Did you know you and I are both really monkeys?

Hominid traits

Genetic pool transfer

Panama research

Albinism and the howler monkey key

If you wish to question the Divine, just imagine what immense power there must have been for a single bacteria to be able to copy rays of light or shimmers of various colors off water molecules. Color therefore in humans should be known as an extension of God's hand on earth.

Overview Linguistics

It is time to create awareness that it was the animal kingdom that gave humans the phonetic developments necessary for language.

For years our valued academics looked to primates for first dialogues. They even attempted to create a dictionary of first sounds of animals. The problem is they could not think outside the box because something was always horribly missing. Firstly without knowing our academia created confusion on the various sorts of early proto-human types that existed, I call this our missing connectivity link. Then this research paper presented saturation point and clans which in the past without this connection we could not connect the dots of regional words to movement of people's dialogues out of our ancestral Africa. Today the most delicate topic of linguistics will be dealt with in the next few chapters. A topic no differently dramatizing for some then the race portion presented. Language officially has been the biggest taboo of earth since the birth of Christ. Which will be an excellent precedent needed for series two. The good news is awareness is the difference between walking in the dark versus walking in the dark with a flashlight. As human species, the developmental stages of language was specifically put in this book as a communication aid for the future, since we are evolving.

Criteria's For Research Chapters xxxxx skim it

> The study on animal social structures to that of observations in linguistic development

  - Such as:

    1. Social rank in monkeys;

    2. The animal kingdoms most required items like water, or food; and

    3. Reactionary noises to fear

  Location of study - Barbados inland specifically Howler Monkeys, Grenada's Rain Forest, Panama Rain Forest and countless hours of National Geography videos.

> The analysis of North America's Birtur descriptives words (Landmass 2) and our attempts to organize them based on their sound categories, done in a sampler format.

> The search for dominant sounds in words.

To understand language we have to find clues that we have systematically build how language was developed starting from its point of origin, referencing animals to our earliest humans. Later we will analyse that different languages had to have come to earth at different periods in time. Our focus will only be a few major languages and predominant language families.

Let's Begin

For this chapter unlearn everything you have contained in you. Before radicalism sets in lets understand the political implications of language today. Bear in mind the previous chapters topics on our One Family Theory, Clans, Saturation Point, the timeline known as the stagnant period of life population's numbers and the dispersed communities of approximately 500 families between Landmass 1 and 2, at a minimum of million years. Yes we are all related and it's about time someone brought this out since we need some type of unity to tackle major environmental problems today. What does human connection have to do with language? Like our first Mona Lisa painting of the world when she was born, there was actually a first dialogue between the species that made her. Our first human(s) mutant.

We are now going to attempt or recreate the puzzle of the unwritten portion of our prehistory. I call this the missing million year gap. Today language has strong national attachments or associations to bordered areas, please erase these connection as well. As we head in to a new Century we have the technological advancements, logic, rational to separate now the world of politics with that of academia.

Our main focus right now is:

1. To exclude politics from our global history;

2. Evidence what topography and animals have given humans in terms of our first dialogues; and

3. Understand what the quantitative component of time is.

Research Methodology

Starting with historical. There is a mention in the Bible which indicates that there existed on this planet one language before it punished humans for being too ambitious and created many distinct languages. As a researcher what if I told you this account of Genesis 11:1-9 was possibly true, would you believe me? This is the actual quote from The Bible:

When the whole earth had one language made up of only several spoken words. *"and the Lord said, "behold, they are one people, and they have all one language; and this is only the beginning of what they will do; and nothing that they propose to do will now be impossible for them. Come, let us go down there confuse their language, that they may not understand one another's speech" so the Lord scatter them abroad from there over the face of the earth, they left off building the city. Therefore its name was called Babel because there the Lord confused the language of all earth; and from there the Lord scattered them abroad over the face of all the earth." (7) From* Babylon.

In this small paragraph something is missing, which is Babylon ancient historical name. In the cradle of civilization, it is known as Babi-ilim which translated means the door of knowledge.

❖

On route to the Toronto Ontario Museum to visit the Mesopotamian exhibit I got reminded again of Babylon's historic name which is Babi-ilim (door to knowledge), that is odd I would always say since that has its origins in proto-Turkic. The exhibits are laid out beautifully and as I stare in amazement another one catches my eye and this one reads *"in the land of Shinar"*, that is funny too since that is awfully close to Cinar in Turkic where in the Turkish alphabet the C is pronounced with an SH, which if translated would make it part of a descriptive word for, the land of Pine Trees.

I am in the second half of the exhibit and there is this beautiful gold goat with feathers called Ur. I now giggle that is what we pronounce as the word Ogur which means luck and sounds exactly the same, just written differently. I am starting to notice Goats, feathers, birds which represent afterlife or deities and that also seems to have an unusual common pattern from an Anatolian archaeological perspective.

By the end of the exhibit I am thinking about the similarities with that of my ongoing research on clans and how much I am convinced our historians may have missed something. Keeping in mine I am not a person who specializes in any of these fields of historical academics but notices phonetic similarities that was taught to me when I took becoming an ESL teacher courses. As I research Babi-illim further I am wondering why we continuously use its fabricated name of Babylon and it linguistic origins (real name) are not really accounted for in any book I have looked at.

After countless hours at the University of Toronto, I start noticing other patterns. The old original historic names of towns along coastal lines or tributaries, numerous mountains, deserts, and a handful of countries also seem to have linguistic similarities. I continue investigating but what's more remarkable is that, they are in a language I understand, it must be in a form of proto Turkic. I am flabbergasted because these patterns, literally litter many historic regional coastal lines and in some regions show clear dominance.

I investigate some more and interestingly the names are very descriptive and reflect Pictionary diagrams to locations. Is this a coincidence? It starts with investigating our prehistory language. My research expands here and will eventually illustrate how the radicalization of language was developed.

The saddest irony is for centuries we have attempted to study our historical past digging deep in the strata layers of earth or by analysing carbon dated objects left behind. But did we miss the most important historical piece of evidence out there. This is the global categorizing in patterns of phonetics and its connection to the study of the chronological ordering of big language families. Called the study of historical linguistics. All derived from our animal kingdom our study focuses on the stages of language development. Written out in the possibly incremental stages. THERE ARE TABLE REMOVE

Stage 1 - Study of First Sounds

During our study of first sounds the one element that has to be understood is, what seems like gibberish to the average human ear from an animal, would have such distinct tonalities that their dialogue is completely understood between those who are communicating it. The following list is a combination of attempting to put together sounds from animals to those generated by humans. Meaning surveying humans in regions of the world that are first settlements followed by carefully listening to different primates in their own habitats. Then merging a list together in cataloguing our first noises. As a researcher the one thing that has to be understood for section 1-7 is the irregularities that has been missed in our history regarding language. This research is to not only prepare for book two but also open up three critical new areas of study based on the movement of people 1) historical linguistics, 2) auditory analysis of Africa's words and 3) anthropological research after linguistic dissection.

> ➢ The mimicking of other animal noises;
> ➢ Different varieties of screeches and grunts;
> ➢ Clicking noises/ plates in the mouth, in the hunt for game;
> ➢ Pounding on lips to generate accelerated vocals; and
> ➢ Ululation - the call of first humans.

Stage 2 - Early Dialogues and Labelling

This area is very critical it is taking stage 1, and seeing continuity of the transformation of sounds to words. The use of same sounds and generating it to words analogy. Better defined as sound to new labelling is the start of early dialogues.

Let's have a look at possibly the oldest word based on primate structures in the world.

Ah (spelt aga) is found globally.

Based on Primate Social Structure

Following Darwin's incremental modification to end result of human,

let's do this linguistically fix

Aa/ aag /aggah/ aga(Pronounced Ah) /aga'm/ human alteration, now in English to the

word "my Leader"

60million years ⟶ 10,000 BC

See this picture, now think of incremental steps of language. This is one of the biggest fallacy of Darwin's Chart a sense of time has not really been understood in these incremental changes. Keeping in mind the English language also has ah or agh - (The thing that doctors make you do). The h in English ah, is equivalent to shaman Turkic's transformation of language, written in Turkish as silent a G. I call this one example of the movement and gradual alteration of words globally.

This is an excellent example of the transformation of some words. For the pattern should be as follows:

Movement / tracking through auditory analysis / originating out of Africa. Auditory video will follow.

Stage 3 - The Introduction of Birtur

Birtur falls under a completely new language category. I am about to introduce the world of Pictionary descriptives belonging to the family of "proto-Turkic languages". Knowing this as the earliest of all linguistics can help us to assess words. As a researcher if I did anything of value I would like Pictionary descriptives to be known as belonging to that of our Shaman History, born out of Africa. This is when the world had no borders. Neglected or ignored by the politics of many countries, or our confusion of early human proto-types and race, it is a wonderful start on our journey to progress.

As we merge my knowledge of Anatolia clan's structure with their linguistics we have opened a whole new look into our earliest world of hominids, especially if a multi-faceted team is used internationally.

Wait am I suggesting "the Turkic family of languages" as a linguistic belongs to that of animal dialogues, absolutely not! That was for marketing, wink!

What I am stating are two things the first is the noises that got transferred from the animal kingdom and their environments to hominids are the base of all phonetics, for all the languages we have today. Secondly language was a million years of building. Shaman linguistics will be evidenced as a relative constant dialogue due to its extremely low population numbers. Known as a language first belonging to the agglutinative family, the sound developments was the starter phonetics belonging to all language categories of this planet. It is the natural steps in the development of actual words and then more complex dialogues, followed by intentional linguistic alterations.

In this section we are attempting to illustrate that language was an incremental process that took a million years and many alterations to get to the 7000 of languages we have globally today.

Stage 4 - The Constant of Language

Studying historical population numbers lets us know that once upon a time our shaman ancestors had one relatively uniform language. Lets unravel our heads from countries for a moment. When we think of the United States with 300 million people predominantly speaking English we can then visualize 55,000 people speaking a early form of proto-Turkic. Here is the catch and it should be clarified immediately on the get go. There are major linguistic alterations that were done between indigenous Turkic and Anatolian Turkish. Turkish although belonging to the same family as Turkic was modified heavily in 1929 after WW1, and unfortunately can cause confusion. Today we will reference the Ulas family continuously to understand our world during the stagnant period of life.

The Ulas Family

The Ulas family clan through inter marriage produced several children who walks on hands and feet, making them a family totalling 19 people in all. Academics were fascinated and called this reverse evolution. They brought in researchers from around the world to study this family, they studied them for several months but did not realize or pay attention to the most critical component of this research, the study of language.

Meaning linguistically this families language remained constant, the only thing changed is that they had developed an inherited recessive gene through intermarriage. This was no different than the events that took place naturally in our environments 2 million years ago.

o   Key Point - The Ulas family case study on clan linguistics, demonstrates that communication may not have been varied but rather incrementally built.

It also demonstrates something altered language change.

In series two this is so critical when we study the dynamics of population explosions.

Stage 5- Overlay

Overlay is simply defined as when one language dies and another one comes in. In the previous sections we learnt that not all words in our language are new. Some like ah in English for example was actually derived from the animal kingdom. What does new mean in this case? New is defined as "old in human timeline" for the case of quantitative time measurements, but new in linguistic timeline of how long human or hominid dialogues have existed on earth.

The concept that linguistic alterations went parallel with clan population increases approximately at the start of 20,000 BC. Known as the grey transitioning period for language change globally. This period is slightly different for all the indigenous languages of the America's and will be addressed as well.

Change in this constant evidences that higher intellect combined with wanting to be different for self preservation is why we have different languages today. This category also falls under social patenting and time.

Throughout this section we are constantly attempting to illustrate the strainer theory, words and sounds originating out of to Africa and now to the development of new language. In addition alteration of language occurred because of the following two types of overlay:

➤ Human Intentional, Altered Linguistics

➤ Geographical Isolated, Altered Linguistic

Why is the process of overlay introduction important here, regarding early dialogue. Primarily to evidence the next topic called the Linguistic Evolution and to show the next steps. Our interconnectivity regarding sounds in terms of our earliest linguistic evolution is how the final outcome was achieve as humans reached close to 7000 languages today.

Stage 6- Linguistic Evolution and Culture

Ever wonder how clicking noises got incorporated into language in tribes in Africa, why their language is different or how variances in any of the languages we have today started emerging? The behaviours of our earliest human prehistory is really not that complex. In the case of the clicks, these were merely behavioural habits regarding the importance of the status of hunt. Regardless of what they were hunting, these clan's early behaviours or their calls during hunting, were merely incorporated into language.

From Africa to China or the fifty or more mixtures of North American Native languages out there, the variances in linguistics is completely a different subject all together. It is what I have called an Overlay, followed by Linguistic Evolution. We know overlay is when one language covers another, as the original language dies out. But it doesn't stop there. Linguistic evolution is the grouping of these words to help aid in incremental building.

Take the timeline 10,000 years ago tribe A who defeats tribe B, kills the men, mates with the females and now 2 intermix have developed in the offspring. This was not a problem during the constant period of life when a uniform dialogues existed but if we analyse the constant attacks and our look at Ottoman Language which is 3 languages in one (Arabic, Farsi and Shaman Turkic). It evidences the dynamics of intermixing. This is the more primitive version than that of the linguistic evolution of European French.

Let's take French as another example, the language itself went through an intermix of 3 or more European languages before it became completely altered to form French as we know of today. This process of building of a language is no different than building a skyscraper, where the end result can now be an identity. The catch is understanding where the core material came from to build this skyscraper.

To clarify:

- Overlay - When one language covers another- historically through violent warfare.

- Linguistic evolution - Saturation reached this intermix generates a new language. Australia case study will follow.

- Social Patenting - When the end result of a language gives to us the emergence of groups of people who can now identify with language. Also known as hundreds of years of generational social input, of these groups children, towards dialogue.

Stage 7 - Chronology of Language

Our attempts to chronologically order languages. Is this actually feasible? The answer is yes if we understand clans and our focus being on African language families. What did these people have written wise, a look at first human alphabets. Lastly which language families naturally spread via nomadic movements and which language families spread via early transport. Like boats or caravans. This section is an attempt to timeline.

Mountains that Whispered

Our Shaman History. Their naming regions or using regional word. Mountains that were never changed, when divisional tactics globally were implemented with population increases. A gift they left behind for humanity indicating "we may not have written but we were here". List is systematically ordered, starting out of South Africa.

REFLECT

Now take 5 minutes and visualize:

That approximately 55,000 of our

earliest human predecessors

roamed over a course of 1-2 million years,

with one relatively constant dialogue.

❖

What this really means

in terms of politics, time, perceptions,

and human development.

*This is called the "**Missing Link**" of our*

earliest linguistic ancestry.

"It Is Our World of Shamans"

Bibliography

1. Behaviour Ecology Oxford Journal V20 14 844-855
2. World
3. Royal Society in non human primates 5 Oct 2009 V 304 no. 1533
4. cnn clip youtube page...boy

A babies ability to coo was actually developed in the jungle.

1. For humour my response when presenting this research is "really are you serious?", "in the 21st century bickering, who cares who crossed" Gee.... bring out its DNA and sound or linguistic cross referencing, I call this the 50 year global mega-project. When DNA is collectively pooled and merged with the movement of sounds and words. Analysis of dying languages are also fully collected and pooled together.

To evidence time and the developmental build of language, let's go through an analysis of our first sounds. How they were derived as our earliest primates started developing thought for survival, in merging vocals to behaviours. Where behaviour to sounds have over time become an interchangeable association. One that followed immediately after by social interaction.

There is no question that animals have interaction and dialogue amongst themselves, this has been proven numerous times as we watch how chimps, elephants or whales interact on any National Geographic documentary. Even recorded noises amongst pods of killer whales can have completely different dialect per pod. Well what about a million years ago when the first set of primates reached a few hundred. What would their interactions have been like?

The following section is the building up of language starting with:

The Mimicking

Mimicking is an act that is a natural component in a topographical surroundings. A parrot with a very small sized brain can copy up to 10 words of its owner. The humour here is when we think of larger animal or hominid dialogue let's not have the metaphoric statement "bird brains" and assume communication did not exist.

Mimicking is an intrinsic behaviour coming from the animal kingdom. When it comes to linguistics this early behaviour is carried forward through ever species that has existed. Therefore the start of phonetic development occurred primarily through the process of mimicking auditory noises from surroundings and the animal kingdom. Take for example the Guatemalan boy who was completely deaf of hearing since birth and received auditory implants at seven years old. Within a year even though he had never heard a single component of sound he can attempt to speak through his instinctive behaviours of mimicking. This is what I classify as an innate response in humans, of mirroring everything in our immediate surroundings which came to us from animals the kingdom.

Forget modern concrete cities, the stimulus of human innovations and what is hashed on to us at the primary school level in terms of educational grooming. Known as the A,B,C's of learning of our modern life. Let's envision what a million years would do to early humans in their more natural habitats. Every single noise, no different than the deaf child hearing noise for the first time, will be attempted to be mimicked.

The following chapters will demonstrate a fascinating discovery made, as we introduce to the world an even earlier set of people unknown previously. This is going to be based off the research that is about to be presented. Whether it is the Kuu-ing of a bird or swishing of the wind, nearly every single noise made by both topography and that of animals dialects have been transferred over phonetically to first syllable words in proto-Turkic.

This is only applicable to certain set of words a reminiscence of our past. These sounds to words are a direct correlation of its age, location of where it was first developed followed by commencement of social interactions and nomadic movements.

Screeches and Grunts

At the present moment we do have extensive research on animal dialogues. However I can't imagine any way to study animal linguistics of the past other than what our academics are working on today and trying to do a comparative. In saying that when we look at Charles Darwin's evolution chart followed by a few National Geography documentaries we can deduce the first noises even thought communicable amongst the species themselves would have sounded like grunts to us.

Let visualize, and do comparative with what our valued academics say are our earliest and closest relatives, the monkeys and apes. The very first noises had to have been grunts and screeches. Does this make sense, absolutely lets continue!

Clicking noises/ Plates in the Mouth

Ever wonder why large circular plates existed in the mouth of certain indigenous? Before we talk about plates and its associated clicks the interesting aspect of these plates that extenuate the mouth is they are found in both Africa and Central America. Now that we understand clans and our One Family Theory it is directly correlated because of our unknown Shaman history. Interestingly where did the idea of putting plates in the mouth come from? It was born like most things from the animal kingdom where primates and rodent both store food and other objects in the mouth cavities, going back thousands of years. What do our valued academics say we are linked to animals and here is just one more example.

Furthermore those plates were originally our earliest ancestors attempts to mimic using foreign objects then followed by transforming these clicks made in the hunt for game. Let's go back to Charles Darwin's evolutionary chart because I am only on man 3 now.

If we look at the family of Niger-Congo Languages in the South African regions, the clicks mentioned above have now been incorporated into full language. Still looking at the Darwin's chart, and using his chart as an analogy, this transformation of communication is man 4 and 5 now. The key to understanding language is all these transition from noise to words occurred during the hominid uniformity period of close to a million years. Over the course of this period we can attribute dialogue to what is known as the development of interaction and social behaviours, in our larger first human proto-types.

Pounding on lips to generate accelerated vocals

Every watch cowboys and Indians? Remember the pounding on the lips to generate noise. This is not a new trait and is tractable back to Africa as well. In my case questioning people of African descent. Based on what they indicated, upon any type of excitement the pounding of the lips are done. We looked at race and have historically always been confused by the black, white and yellow concept of people. Today we have a better understanding that many African communities picked this habit from their earliest ancestry the African Shaman Natives. Who migrated out of Africa into the America's thousands of years ago and eventually evolved topographically.

Ululations

Ululations are very interesting on the other hand. These are the high pitch tongue thrill noises that is used in most celebrations across Africa, Anatolia and the Middle East today. What is the significance of this odd type of noise is also the transformation it went through. From our earliest ancestors calling each other in its nomadic format, to settlements and now used in celebrations. This process was done over the thousands of years of social interactions.

Interesting Fact - Why don't the Natives have ululation the way Africa and the Middle East has it. The hypothesis is because of human's adjustments to topography. The harshest climates with the coldest temperatures falling below -50 degrees are regions of the world like Siberia, East Russia and Central Asia. Climatic conditions would have most probably made them dropped this practice pre-entry.

This is very important as an aide in understanding social patenting. Especially those pertaining to extremely difficult topographies, such as dangerous rain forest or extreme cold weather conditions. It gives humans a better understanding of time. The pygmies of the rain forest and Siberia's indigenous are an astonishing study of early humans adjustments to extreme temperatures.

Ululations is what I call our earliest component of dialogues, that helped to accelerate speech. Why am I saying component of speech. The complexity of the groupings of the words la-la-la-la in a rapid sequential format (uses different vocals for speech), this is my ESL training now. That la-la-la sequential format meant complex dialogue had to have existed if we analyse this tongue thrill.

In attempting to put a sequential order of some of the first languages of earth section a lot of analysis was put into sounds to words categories. Take for example the process of screech noises to ululations. Which gradually got transformed to sound more like the word "LA" itself and actually belongs to the Indo-European category of first words. Yes, don't be surprised Europe's earliest words were born, as well out of Africa. Meaning if Aga(ah) belonged to the agglutinative family, the transformed high pitched thrill noises to the word LA belongs to the start of Indo-European family of linguistics. The transformation of ululation from screeches was a developmental process, and is related to our earliest settlements. Let's remember this in the chronology of languages section, where there will be more information evidencing this.

The newly discovered Gobekli Tepe in Turkey, is a structure that demonstrate very complex communication had to have existed. From a language perspectives this discovery now changes the entire dynamics of our linguistic history. Why because the structure is not just piles of rocks, it is part of our earliest architecture. Leather pulled material covering Gobekli Tepe dome, huddling with large domestic animals for warmth, and ululations are evidencing the start of our earliest settlements.

Imagine we are on a mountain by a megalithic structure like Gobekli tepe in southern Turkey, our earliest ancestors would ululate to call for a variety reasons. Now visualize regions in Siberia would the same call be done, where the weather conditions are so detrimental one uncovered location on exposed skin can actually freeze and break off?

The category of first sounds from a historical perspective has very little traceable records. Regarding our animal kingdom derived dialogues, it concludes here.

Our First Attempts at Dialogues

The question then becomes when did first attempts of varieties of sounds convert over, to basic dialogue. Clearly there had to be a transformation period to that of words that we can recognize today. When we look at history therefore we are not looking for the sounds and screeches our earliest primates did. We are trying to do an auditory analysis of words from what seems to be a noticeably dominant language, in its proto format once highlighted. These are words because of my personal background that I recognized globally in different parts of the world.

Our quest begins with attempting to categorize these first words. I call this the "*search for words, of antiquity*". No different than that of any archaeologist digging. Words left behind to us from our earliest shamans who first migrated out of Africa. Our earliest descendents who fine tuned these noises to actual words.

Where are their words globally to back my research up, is what starts our exploration. There are several formats as demonstrated in the criteria for research section at the beginning of the book but other than that, what else can humans do to pull out our earliest dialogues?

Indigenous of America

One suggestion is to have a peak into the no outside contact indigenous tribes of America's. They are a great start. In Panama for example because of time constraints I was only able to deal with the more publicly open tribes like the Emberra and Kuna yala. However knowing what we know today are research can focus on these isolated indigenous more carefully. After all the focus is a plate used for noise in the mouth in Africa to a plate in the mouth in the America's, is far from coincidental. And signifies social decor. Remembering that social decor and language had to have gone hand and hand.

To the point that these tribes, their dialogues, cultures, beliefs is a peek into what life was like 100,000 years ago and should be compared from an anthropological perspective to our origins out of Africa. Amusingly said, no time machine needed here.

The other interesting aspect of these all tribes in the America's is that they are all human species and not one is a hominids. Today other than some sight-seeing in Russia we have no records of our earlier living human proto-types. Which brings up another interesting question, how much language did hominids actually have? Historically majority of academics wrote them off as apes. Yes book two will illustrate that there might have been a darker side of our history and why certain things were done. But today we can focus on this transitional period as we look for attempts at first dialogue.

The key is always working backwards. We are told the first humans cross the Bering Strait 50,000 years ago, we can deduce they crossed over with quite a complex language. This is if anyone can understand or visualize the extreme topography of the Bering, to the act of actually attempting to cross. Massive social organization was needed. As a researcher the goal is to trace back as far as we can go. One of the goals in the linguistic section, is to find the words belonging to this transitional period between noises to first words and illustrate it.

Early Dialogue

Ever play Lego and attach one piece to another, the analogy is early dialogues were the first 4 pieces glued together in any of your play sculptures.

The format for early dialogue came through the intermixing of the development of first sounds, its continuous use followed by its gradual refinements in the actual words. The idea of long accentuated noises, screeches or grunts transformed itself to a shortcut in verbal output. For example mimicking the dangers of snakes in the water siiiiii noise became si. Which is an early proto-Turkic, shamanic word for water.

The hypothesis therefore is these initial words started with the birth of social behaviours in hominids. Going to siiiii watering locations, would be an excellent example. This is called association of the animal kingdoms labelling to the birth of actual words. Animals did not care it was water, they cared that siiii was in the water or the association it had generated. This will be discussed in later chapters as something called Pictionary labelling.

The field of socialism has always been highlighted predominantly for the human world but the transitional period between first primate to first hominid also had to have had complex social behaviours. The query is what led to advance dialogues as social interactions started developing in our hominids. The answer is the need for heat, facial and body decor for mating, use of tools and the organization needed in the hunt for game. Meaning our earliest dialogues as we know of today vocabulary wise started with the need to interact and innate behaviours for self preservation. The primary interaction therefore which was the catalyst for organized early dialogues may have been the hunt for game. kurds are important no?

Hunting, the need in attacking game and feeding one's self should be the first evidence of word building as we know of today. If we look at animals some may argue this does not necessary mean early dialogues started during this period. When we reference early dialogues let understand we are constantly referencing dialogue as we know of today. That grey area, that converted the siiii to si. We are looking when si water, became an actual word. In fairness to some academics who will argue this is not accurate, we can leave the act of hunting as a grey area for our earliest humans as maybe having first dialogues. The need for heat however is my belief that phonetic sounds, early dialogues and shaman era linguistics as we know of today had existed at this point.

Consequently when would humans or its predecessors greatest technical advancements have come in to really push dialogue. It would have been the discovery of controlled fire. According to J.M Roberts book on prehistory and first civilizations he illustrates " that many scholars based on newer evidence in the Transvaal area believe that hominids were using fire well before homo sapiens even came into the picture".

If we were to debate this, then "many scholars", is still not a uniform agreement to suggest language occurred over a million years ago. But I am on the side of the scholars who do believe fire and early dialogues came approximately the same time, and that it existed way before the existence of homo sapiens and here is why. Like all living things that were born and evolved incrementally from micro organisms to large mammals, so did the languages of earth. Significantly language was a developmental process in our history. And unlike popular belief was not born with Sanskrit or any of the Mesopotamian languages. It was born in Africa.

Time Chart (4 million years to 100,000 BC).

4,000,000 BC Appearance Australopithecus	3,000,000 BC	2,000,000 BC Homo Habilis (Tools)	1,000,000 BC Homo Erectus (Fire)	600,000 BC Neanderthal Denisovan	100,000 BC - onwards Homo Sapiens complex language

Our Shaman Era———————————➤

The colour coding for the boxes Homo Habilis and Homo Erectus are shaded in the same because I would like to add another interesting fact as we try to pinpoint and evidence the undocumented portion of our language origins. It is also the reason the arrow is not fully on the Homo erectus family.

We mentioned that some scholars might argue the need to hunt does not constitute early dialogues. Lets attempt to evidence something different then. Tools were found in Ethiopia which are the oldest to date at about two and half million years. They are called pebble choppers which are all selectively prepared. As basic as these tools seem someone/ groups of people from our past took the time to carefully chip away at these stones in a very thought out manner. Very different from certain animals in the animal kingdom who randomly picks up tools to aid with hunting or cracking food. Some will argue that such exciting evidences by academics as fire and tools are not sufficient when it comes to evidencing language.

I beg to differ. For one thing animals have dialogue. But the thought to create something so precise for hunt at about two half million years ago means we should really be investigating on how we can evidence when that animal dialogues converted to words we know of today. Known as the sound to dialogue transitional period, with the focus always being around the shaded areas in the above charts. This is in periods where conversional words are born. This is like the word aga(ah) in the previous chapter which ironically probably goes as far back as two million years to the time of Ethiopia. The title of the books suggest our animal ancestry, well aga(ah) from the previous chapter, is a perfect example of the transfer of words over time. As limited as these words may be.

The following chart is a look at sounds to early dialogues and their possible scenarios. All of these words are Shamanic phonetics because they are noises or environmental sounds made to words and socially still accepted today. Interestingly they are all found originating out of Africa.

Birth of Early Dialogues Derived from First Sounds

Phonetic Sounds	Descriptive Source Animals	First incorporated into proto Turkic, these are some examples	Translation English	Feasible "Cause and Effect" Possible Scenarios
Agggg	Monkey	Aga	Leader	Hierarchy in Animal
Gahh	Earliest hominid	GaGah -si	Beak	Beak by water
Siiiii (Same as sea in English)	Hissing of a snake	Si / dialect transformation to Su	Water	Animal recognition and safety, not the word of water itself-They don't have capacity to label and really care. Meaning word was born with no developed cognitive thought.

Beuh	Monkey	Bocek- cek is the suffix added - Reactiona	Bug / something scary	Reactionary Fear
Eeeee	A repetitive e Monkey / Mammals	Yilan- lan is the suffix added	Snake	Reactionary Fear
Kuu	Bird	Kulu - K was added on after Islam. (Ulu is powerful mountain)	Servant of God	Mimicking Behaviour
Ka listen to forest video	Smaller mammals	Kawa / Native Word	Proto Turkic Head	Mimicking Behaviour
Ürü	Bird	ürün	Product	Mimicking Behaviour
Uuu	Bird	Ulu	Big and powerful	Mimicking Behaviour

Words are not limited to where they can be derived from. Meaning the transmitting of noises back and forth from just the animal kingdom is not exclusive. Then there was the topography related or heard noises, made to words.

Phonetic Sounds	Descriptive Source	First incorporated into proto Turkic, these are some examples	Translation English	Feasible "Cause and Effect" Scenarios
Ush (accentuated sh)	Topography related	Uş (Ush)	Bird	Sound of wind
çat / Chat (accentuated t)	Weather	Chatt	The one that strikes	Lightening
Aw (loud vocal)	Sudden reactionary noise	Hawa	Sky	Sky related

Early dialogues give us an understanding that the formation of social structures in the animal kingdom meant there had to have been understandable and advance verbiage in putting in order of ranks within a community of species.

We know early dialogues amongst predecessors of humans up to date has been an unidentifiable greyer area of research. This is because our research methodology were only really limited to artefacts regarding language analysis.

We looked at the animal kingdom and our environments which gave us first sounds, and how it has aided us in building first dialogues.

When we look at other languages there are definitely more unusual categories and so many other languages, that a cloud of smoke may appear regarding linguistic development.

The truth is humans have not stop building our ever growing linguistics. Mouse, computer and screen are words that were just incorporated in the last twenty years. This small scale injection is no different than what our first settlements gave 10,000 years ago. Domination of territory and the comfort of settlements are the real reason enlightenment and language merged.

Let's play a game to understand what development really is!

Lets envision Buckingham Palace and now have a look at the diagram below. Can you see the disparity of the early stages of what is to be a foundation of the palace, to that of the development of language. I am attempting to use an analogy of how things were built

Significance of our earliest ancestors need for heat and actually socializing it

Let's look at prefix, suffix or conjoining example of words belonging to agglutinative families. I have picked a random example found in Birtur descriptive words the list is at, bat, yat, hat, kat.

Ten thousand years ago with the full start of indo European languages you can now find a similar order in a completely different language "English", at, hat, cat, bat, tat.

What is the difference between the two categories 1) age 2) dialect 3) pronunciation.

What would be the commonality? It would be the transfer of certain sounds between the language families. By studying regional words predominantly in Land mass two, we can attempt to categorize certain base sounds that got incorporated into words from linguistic in Africa. Keeping in mind we are limited due to no audio equipment and time restraints, this research is just an overview using very common category of words and sounds.

The table below with words thrown in does not illustrate well the language of our earliest predecessors, therefore I will label it a basic sampler. It gives us an overview of common sounds from our surroundings. Since these two tables are referencing vocal sounds to the shift towards first syllable words, an audio webpage and graphics will follow. It's objective is to demonstrate surrounding sounds to how word formations were born.

The introduction of Birtur and our Pictionary descriptive index in the back of the publication along with nomadic patterns, graphs and diagrams in later chapters will substantiate this further.

When we study history let's remember it is not only artefacts that talk to us!

Before we commence this chapter. Let's do a sneak preview for series two. The reason is our animals and hominids did speak, the key is evidencing this without the politics, or offence or radicalization attached to language. Language is a very touchy topic, the World Wars brought on the need to dominate lands, carve out borders and bring out national flags. This is an innate human behaviour or response to self preservation. Technically we are part of the animal world and language is merely attached to territory no different than the analogy of the Lion's control of his kingdom. This book is a preparatory to book two and its clear objective is to break perceptions. What is Pictionary Descriptive? It is defined as language classification that I have labelled as Birtur. It will illustrate our world of dialogue for our first shamans of earth, inclusive of our earlier hominids. Meaning no we did not jump from apes to settlements more humorously said. We developed very slowly, and incrementally in a ecosystem over the course of thousands of years. Pictionary descriptive therefore should be classified as a linguistic artefact, it is a huge area unexplored for anthropological research.

Introduction of Birtur

Close to a million years ago hominids discovered fire, the hypothesis to follow is that it is around this period that the start of conjoining, single syllable words, started forming. These are the continuity of same sounds converted from facial expressions, to surrounding events or objects. Followed by the start of single syllables. These are called reactionary responses. Therefore what are visually associated phonetic noises? They are the commencement of the standardization of sounds as a component, of starting of speech. And lastly to its slightly more advanced version which are full detailed descriptive words, known as "Pictionary visuals" of the first words of earth.

For the purpose of this publication I will call this particular category Pictionary linguistic, Birtur. Its name is translated in Turkic as *"first type"* and these newly discovered patterns should be dated as an earlier version of proto-Turkic belonging to our *"Shaman History Era"*. Known as the dialogues of the 55,000 early human proto-type, who roamed for a period close to a million years. Their language belongs to the proto- agglutinative family.

Also described as the language of the First Clans that grew and broke off, from their original pods of a few hundred hominids somewhere within the African continent. It is our attempt now to classify and organize these first word visuals which will also help in tracing the migration of our earliest nomads.

Other Characteristic

Birtur is the foundation of all languages, born predominantly to the proto-Turkic family of languages originating out of Africa. That transposed itself in terms of first sounds of 1 relatively constant dialogue to the 6,909 languages we have on earth today. In other words what the Greeks gave to the world as enlightenment, is what Birtur gave in terms of the first vocals and syllables, from our environment to the world of languages we have today.

How can I be so sure? When we look at linguistics we should also look at its evolutionary stages, and its complexity levels.

Birtur words started off as Pictionary Descriptive for one thing. Secondly they are logical in order regarding lexicon structure, this was done when the world had a stagnant population and no real need for the implementation of distinctions. I would be referencing linguistic distinctions in this particular case. The need for human distinction, I am different from you ideology is especially important in understanding for series two.

language that is agglutinative in structure and is vey simple phonetically references the following, that what ever is sounded out, eventually became exactly its equivalent in writen format. When we look at English for example its complexities evidence highly developed "human altered" language modifications to its predecessor the agglutinative linguisitc structures.

Advanced gramatical formats such as those pretaining to the any of the Indo-European family of languages (English, German, Farsi, Greek, French, Armenian) confirms that these language belonged to grammatical structures that came way after. It would also clarify why all Proto Indo-European languages were once agglutinative in structure and one of Semitic's proto format for example, Akkadian, also has many Birtur terminology in its language. A key in understanding the order of human language development.

Interesting fact - If you go into a forest, almost every sound in our environment today that a human can mimic, can be found as a phonetic word in Turkic. This is due to its age.

Here is an example list of the first noises derived from the start of those descriptive words.

For the following reasons:

1. These are Pictionary starter words, what our environments contributed in terms of development of language in early humans;

2. There is a direct correlation between these descriptive words to immediate surroundings and regional names, or actions;

3. These are single syllable or very basic conjoining words;

4. Some are decoded words phonetically, but not necessarily found in its same format in modern Turkish or Turkic;

5. They may have transferred over linguistically but because of its age they may have different meanings to that of the Turkic family of languages today;

6. They are consistent patterns or grouping of words that are associated with absolutely "*no archaeological find, or written portion*" of history but rather the study of language;

7. And lastly for educational purposes. To illustrate its phonetics, word order and linguistic contributions to a variety of languages we have today.

Birtur Classification Criteria's

What will be our methodology? When a research document is presented it is one thing to let readers know that these could be the first words but it is another to try to implement the actual illustration of it. In the end the biggest challenges comes from evidencing to readers our unwritten portion of our history. It is exactly why this categorization of words in our index page that will aide us with the necessary back up, that an irregularity exists. A pulling out of another set of people.

We know after extensive research that Birtur is what our first hominids saw topographically and its association to the development of language. We have also found these regional markings of its proto Turkic format predominantly along coastal lines, mountains / desert ranges and some islands. Which gives us the clue and helps us track their complex linguistic of our earliest nomads and their preferred migration patterns. Before we go into the analysis of actual human language have a look at the following diagram. It is our visual aide to understand the difference of a nomadic makeshift settlements to that an actual settlement. The start of settlements was the catalyst to accelerate language change globally.

Visualization Exercise

		Understanding the Origins of language:
Remember!		Newly discovered Gobekli Tepe's age is carbon dated to approximately ck 15,000 BC. It is an example of nomadic patterns going back to a point of origin / settlements. According to scholars however controlled fires are over 1 million years old. *This is part of their indigenous territory.*

Now Let's Play Pictionary

The Oxford definition of the word Language, is the method of communicating in a structured or conventional way. It is so important it even contains the shaman indigenous word Gua which means prayer.

We can deduce by looking at the Ulas scenario one critical factor in the formation of other languages, is their mutation did not alter language.

Having attempted to put together language order and researched the chronology of linguistics with knowing the answer. The question I pose to readers just to provoke some thought is why in the last 15,000 years were there a drastic changes in linguistic development.

Animal to Hominid is exactly this. It is a precedent for awareness on all frontiers and our attempts to understand linguistic evolution. The answers regarding radical changes will follow but here is a clue, the same reason humans went from a stable mark of 55,000 people over a course of a million years, to 7 billion people today.

Professors over the course of many years have attempted to categorize and create dictionaries of the dialogues of various primates we have today. Have we missed the point? This type of research is great when studying animal dialogues but we have evolved into a different species altogether. Understanding the key element here is we all evolve out of a previous species, so our tendencies at looking at the animal kingdom for an answer is normal. The problem is looking at apes, orangutans or chimps for early human linguistics and vocalization may not be the best form of abstraction regarding the study of language.

Keeping in mind time, there is a 2 to 4 million year variance between animals that have simply always existed with humans versus our actual predecessors like a Neanderthal or a Denisovan. The time icon is there to highlight that we are missing over a million years of linguistic history because those species do not exist today.

Remember Marco Polo and his travels across Asia and the Indian subcontinent.

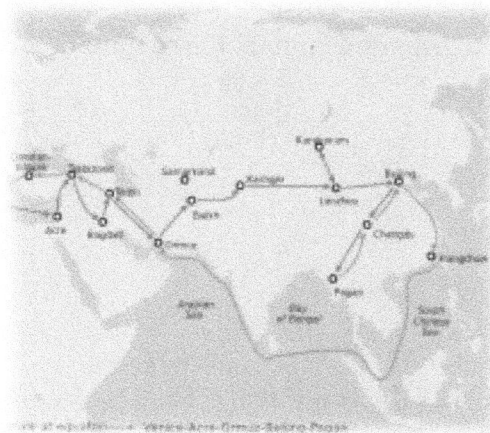

In the 1300's with his basic Turkic he was able to travel and communicate with many of the Asiatic tribes that had varying degrees of Turkic dialect without too much difficulty. No different than how our nomadic clans travelled 1 million years ago as they started developing their own regional dialects, which occurred historically. The simplicity, logical order and the ease of noises to word association evidence our theory further.

The question I pose is what occurred historically that brought on such radical changes in language? Now let's debate?

The next paragraph is called the "first language" pondering, these dual perspectives through the publication is to aide us on being more tolerant on the topics presented.

In the previous chapter we had a look at how the Ulas Family's language remained constant, even during a recessive gene alteration on the part of their children. Historically academics on both the linguistics and anthropology side have written off majority of the events during the Palaeolithic period because of its ambiguity. But an examination behind the idea of the transfer of language may help us decode sequential events. The question then becomes when did the constant of language change over? How can we determine on our human timeline an approximate time frame to that of not only basic dialogue from the animal kingdom, but a complete linguistic alterations globally to what we have today?

x

Using unconventional methods and the concept of what is a linguistic constant or the transfer of language (the casestudy on the Ulas family), it will help guide us to chronologically organize the major language families we have today. When we think of early humans the first train of thought should be when did language fully become language? Not the reciprocal opposite of comparing it to the animal kingdom, we already knew animals had language.

Meaning when did a language occurred that we can understand at this present moment. When we look at the Oxford definition of the word language the definition above in the introduction should be modified as follows "*the method of communicating in a structured or conventional way, that humans can understand effective today, from our earliest prehistory*". Oxford should also have two meanings one is for animal language the other is for human language.

This is what I define as the real meaning of the word language. So our search commences here, we are looking for which language base transferred from our earliest predecessors that humans can understand as of today.

Building of Language

The following diagram contrast the stagnant period of time of clans which is 55,000 people over the course of a million years, to that of building of first words. I call this the domino effect which is the transferring process of information from one generation to the next. No different than the case study of the clans who adjusted to a dangerous tropical rain forest environment, this is thousands years of the transfer of information and building of first words. graph

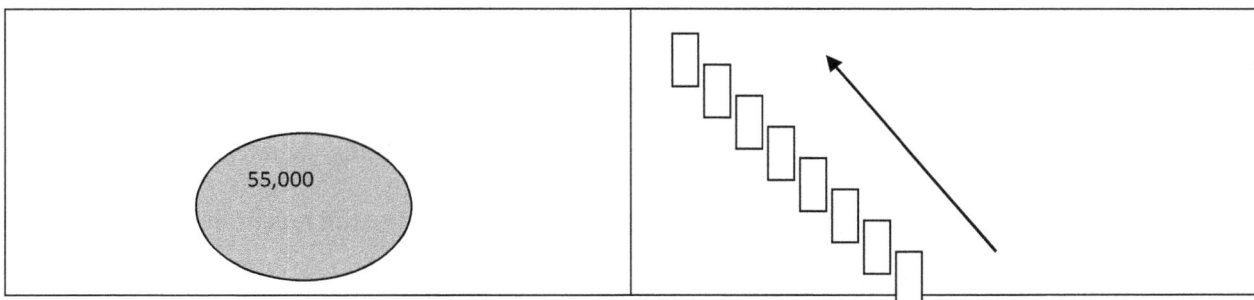

Stagnant period of time for clans and the importance of understanding historical population numbers when it comes to our prehistory on word building.	
	Domino effect and growth of words

The question then becomes where on our human prehistory timeline can we determine not only when human dialogues were born (with today's understandable standards) but the timeframe of when it commenced.

I believe the answer is more unusual then some can imagine. I merely researched when humans or its predecessors greatest technical advances came in, which would have been the discovery of controlled fire. According to J.M Roberts book on prehistory and first civilizations he illustrates " that many scholars based on newer evidence in the Transvaal area believe that hominids were using fire well before homo sapiens even came into the picture".

If we were to debate this, two ideas come to mind the first is, what do you mean that early predecessor to humans were using fire? Were they not under the label of animal? Secondly that "many scholars", is not a uniform agreement to suggest language occurred over a million years ago. We are always referencing a connection of a linguistics that we can understand today.

The anomaly here is, there is a major division that exists amongst academics. Some indicate a million years ago there were only expressive noises and grunts, while others indicate there may have been complex language but we don't have records of it. I believe the research behind Animal to Hominid has surpassed both of these academic trains of thought.

When we look at what is a controlled fire, I am on the side of the scholars who do believe fire and language came approximately the same time or possibly even earlier. That the thought behind the creation of smoke on our historical timeline meant language had to have existed way before the existence of homo sapiens and here is why.

Like all living things that were born and evolved incrementally from micro organisms to large mammals, so did the languages of earth. Significantly the building of language was also a developmental process in our history.

Even variations in vocal repertoire, amongst first animal proto types, I am referencing the period earlier than 2 million years had to have been understandable amongst the groups that communicated between them. It is just unfortunate that we as humans don't have records or understand the various tonalities of vocals of primates today that each noise, screech or grunt signifies.

In Animal to Hominid the objective is not that a million years ago there was no language; the criteria for research becomes can we timeline or evidence it with nothing on hand? Going back to J.M Roberts book on prehistory I have taken the dates of the various species that have existed which is our earliest ancestry and I have added a variation to his timeline. I have added this transfer of knowledge or building of words so to speak, between species.

Our new and improve created timeline illustrates where language transferred over from communicable grunts to full dialogue in humans. I call this the communication that is understandable in the 21st century using the simplest probability when did controlled fires begin?

Time Chart (4 million years to 100,000 BC).

4,000,000 BC	3,000,000 BC	2,000,000 BC	1,000,000 BC	600,000 BC	100,000 BC -
Appearance Australopithecus		Homo Habilis (Tools)	Homo Erectus (Fire)	Neanderthal	onwards Homo Sapiens complex language

Why are two time periods shaded in well there is another probability. Our analysis of the birth of language may not have just originated with fire and this is why the colour coding for the boxes between Homo Habilis and Homo Erectus are shaded in the same. I call this the probability zone of when understandable dialogue first emerge. I am attempting to evidence two things here regarding systems of communication and our prehistory time line.

The first is there has always been language in the animal kingdom - we have passed that awareness and secondly this timeline simply illustrates the developmental process of thought to that of language. With the shaded in areas being the start of more complex adjoining single syllable words.

We know fires are very well thought out, the grinding of two pieces of sticks for close to an hour required complex dialogue but more interestingly tools were found in Ethiopia. These tools are the oldest to date at about two and half million years. They are called pebble choppers all selectively prepared. As basic as these tools seem someone/ groups of people from our past took the time to carefully chip away at these stones in a very thought out manner. Very different from certain animals in the animal kingdom who randomly picks up tools to aid with hunting. Thought and reasoning is what I call a prerequisite for language. In this case hunt, more precision designed weapons like sharpened rocks, and the vocals to organize the whole process.

Some will argue that such exciting evidences by academics as fire and tools are not sufficient when it comes to evidencing language. I beg to differ. The introduction of Birtur and our Pictionary descriptive index in the back along with nomadic patterns, graphs, and diagrams in later chapters will substantiate this further. In addition, what is better evidence than the Genographic Project DNA Migration Map as our guide anyways?

Landscape 2 isolation known as the America's, gives us:

- Word clues from our earliest ancestry;
- Descriptive details of surroundings;
- The ability of categorizing phonetics;
- Age of language; and
- Dominance of first words to that of an understandable language found today.

Then there are the no outside contact, indigenous tribes of South America's, which are a great look into the past where all their products are not only biodegradable, with constantly changing makeshift settlements but full dialogue amongst the members themselves does exist. To the point that these tribes, their dialogues, cultures, beliefs is a peek into what life was like 50,000 years ago and should be compared to our origins out of Africa.

In the index page at the back of this publication you will get the entire list. It is our attempt to categories these words to illustrate patterns associated with early human dialogues. Here is a list of the first noises derived from the descriptive word study predominantly the Americas or more topographically hostile landscapes such as Siberia for example. We are working backwards from Landmass 2 to give us clues.

If you notice linguistically I have classified Birtur a completely different category to that of proto-Turkic? It was completely intentionally done. Proto-Turkic is a language, an earlier version to that of any of the Turkic families of languages out there. Birtur is a classification of visual descriptives, they house predominantly proto-Turkic words. The other reasons are as follows:

8. These are Pictionary starter words, what our environments gave in terms of sounds to that of language development.

9. They opened up a different kind of archaeological department, this is our historical records based on the study of words;

10. There is a direct correlation between these descriptive words to immediate surroundings and regional names, or actions;

11. These are single syllable or very basic conjoining words;

12. Some are decoded words phonetically, but not necessarily found in its same format in modern Turkish or Turkic family of languages today;

13. They may have transferred over linguistically but because of its age they may have different meanings to that of the Turkic family of languages today. This is a very important element of Birtur;

14. They are consistent patterns or grouping of words that are associated with absolutely *"no archaeological find, or written portion"* of history but rather the study of language;

15. And lastly for educational purposes. To illustrate its phonetics, word order and linguistic contributions to a variety of languages we have today.

The critical importance of the Agglutinative Language family

Why are the family grouping of agglutinative languages so critical in our attempt to chronologically order languages.

Agglutinative languages were the first dominant grouping of languages of earth, and the most naturally spreading. "Most word are formed by joining morphemes together" or easier suffix additions by gluing single syllables. In general the family of agglutinative languages are easier to learn, especially those associated with Birtur, because they were created with sounds to phonetics, incrementally done over time. Remember the concept of knitting in the acronyms page. Agglutinative languages are the yarn to the various outfits that have been produced over time.

These words are naturally developed in our linguistic history and require less memorization. The critical importance of the Turkic family of languages is that everything, absolutely everything said can be written out phonetically regardless of dialects. Remember nursery rhymes? The start of language was the start of basic sounds to the changes of single words, proto Turkic for example has: at, bat, yat, kat, hat. These types of words transferred over to other family groups of linguistics with different sound but same order as follows. In English: at bat cat that and hat

The critical importance of the Agglutinative Language family

Remember! Why are the family grouping of agglutinative languages so critical in our attempt to chronologically order languages.

Agglutinative languages were the first dominant grouping of languages of earth, and the most naturally spreading. "Most word are formed by joining morphemes together" or easier suffix additions by gluing single syllables. In general the family of agglutinative languages are easier to learn, especially those associated with Birtur, because they were created with sounds to phonetics, incrementally done over time. Remember the concept of knitting in the acronyms page. Agglutinative languages are the yarn to the various outfits that have been produced over time.

These words are naturally developed in our linguistic history and require less memorization. The critical importance of the Turkic family of languages is that everything, absolutely everything said can be written out phonetically regardless of dialects. Remember nursery rhymes? The start of language was the start of basic sounds to the changes of single words, proto Turkic for example has: at, bat, yat, kat, hat. These types of words transferred over to other family groups of linguistics with different sound but same order as follows. In English: at bat cat that and hat

Bibliography

- Cognitive science v.24 2000 page 445-475 Marc D Hauser Harvard University

- live science tanya lewis

- oxford dictionary language

Everything, absolutely everything that exists on this earth today came from our environments. Inclusive of thought and the incremental development of language.

there is a major link between africa and the americas the problem is politics or historical race relations clouded everything in between. Today academic teaching via media is making us understand this connection. I call this the start of real progress in humans. Welcome to the information age.

Circle analogy A to B to C-The bubbles represent the transfer of language over course of million years.

A good comparative is our case study at the micro level and the interesting events that went on in Southern Turkey.

shaman word overlay/ integration / coshare word / variance 7000 languages

interpose integration and the borrowing of words

The Case Study: From Animals To Humans Understanding Our Linguistic History(A) Homo Erectus & (B) Homo Sapiens	
Transfer of Languag (A) B C	Based on Primate Social Structure → Agg/ aggh /aggah/ aga /aga'm/ modified now - leader. This is an excellent example of the transformation of some words. Movement/ through auditory analysis/ originating out of africa.

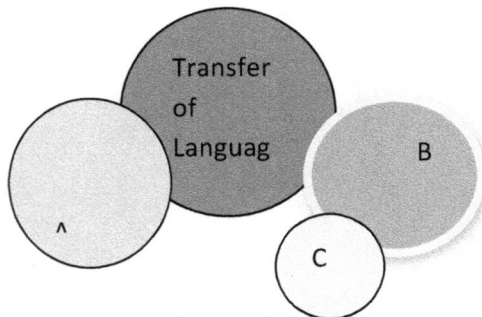

	The transformation of phonetics to modern languages including Turkish took thousands of years. The last 3 early humans belongs to our hominid dialogues. Today less than half of words in xxxx families can be traced back to Africa. the dissection of word in isolate native languages, Africa comparative gives us the clue. Overlay and chronology of language will explain why.

The theory on saturation point and the missing million year gap in our history should go together simultaneously. For a lot of our historical religious figures it is known as the people who practiced paganism.

it is exactly this duration when not only did we start becoming human but when language was born. The question is what prompted change in languages. This analysis is so critical for human progress. This chapter is dedicated to a concept called overlay.

Lastly we have to remember the episodes of our past including hidden or negative images of paganism, is also our history.

notes;

Am I making the planet Turkish no, error Turkish is linked to Turkic but it is a Europeanized -
modified language 1929. even as regions experienced overlay the reminiscence of their shamanic
history remained because of its age.

500 pockets /	7billion
1 general dialogue /	7000 languages rapid natural change

I am merely evidencing for the 500 pockets of settlements for a million year period we have the
gift of proto-Turkic shaman labelling of regions before overlay, changes in skin color and social
patenting took place. same diagram as Charles Darwin but for linguistics and incremental
building from the animal kingdom.

pic

this is perfect timing native =green blah blah why al gore did had no real impact here are some
proposals.

book 2 will illustrate how we derived at 7000 languages today a parallel growth on a chart with
that of exponential population numbers. reproduction cant be implemented if we don't
understand precedent.

Intentional changes in language is equivalent to self preservation threats, defensive strategies, to protect during war, ...book 2

analogy understanding the illnesses of the root problem will aid in how tall the tree can grow.

We know langauge went in the following order.

sound, words, language and lastly variation called overlay.

⭐ **Interesting Fact / Time-** 20,000 years ago when there was no shelter, or lit up roads, when there was only twigs wrapping animal hide onto blistered feet language started advancing.

Visuals of what was going on in our environments started to reappear year after year, receding flood waters, melting of snow, sparkle of water, hunt or barley season. Gradually visual descriptives are the foundation of what made certain calendars. A look at the Jewish calendar months will show its age and Birtur descriptive association. These words therefore are a weaved fabric in socio - cultural customs as religions develop themselves.

The most critical component of Birtur is what our earliest ancestors visually saw and attempted to attach as phonetic sounds to that of their topographic visuals.

Advancing is the understanding that language existed way prior to the start of settlements. language and radicalization is recent. self preservation meant control of land or territory no different than the animal kingdom.

Historic Clan Names indicate a grouping of families	Illustration and Origins of Clan Names
	Native Americans Anasazi Tribe, Israel Yerusha Clan, Etruscan clenar -clan

	Suffix endings were also used to depict large families, clans or tribal groupings: • Armenian ian's (a good example of suffix endings Kardas-ian), • Turkic oglu / oglari, (example Saruhan-oglu) • English brothers (common last name, exactly as is brothers) • Any of the countries with the ending stans that have grown to be countries today came from very large clans. Kazakhstan, Uzbekistan, etc. Average number of children a single women produces is 8-10

Linguistic Evolution

Linguistic evolution is the continuous change over time of how groups of people express and associate themselves in terms of identity.

Birtur is the foundation of all languages, a form of proto-Turkic that transposed itself from 1 to the 6,909 languages we have on earth today. In other words what the Greeks gave to the Europeans, is what Birtur gave in terms of the first vocals and syllables, to the global languages we have today.

How can I be so sure? When we look at linguistics we should also look at its evolutionary stages, and its complexity levels.

Birtur words started off as Pictionary Descriptive for one thing. Secondly they are logical in order regarding lexicom structure, this was done when the world had a stagnant population and no real need for the implementation of distinctions. I would be referencing linguisitc distinctions in this particular case.

A language that is agglutinative in structure and is vey simple phonetically references the following, that what ever is sounded out, eventually became exactly its equivalent in writen format. When we look at English for example its complexities evidence highly developed "human altered" language modifications to its predecessor the agglutinative linguisitc structures.

Advanced gramatical formats such as those pretaining to the any of the Indo-European family of languages (English, German, Farsi, Greek, French, Armenian) confirms that these language belonged to grammatical structures that came way after. It would also clarify why all Proto Indo-

European languages were once agglutinative in structure and one of Semitic's proto format for example, Akkadian, also has many Birtur terminology in its language. A key in understanding the order of human language development.

Take for example the word knife, very misleading for those attempting to learn English which should be prounounced or spelt like nife. In Birtur this word would have never existed for three reasons.

The first being it is not simple enough for primitive humans to produce (not in the language developmental stages which were facially produced expressions of first noises, to that of single syllables), secondly the word knife if we visualize it, is not a descritptive word of that stagnant period and lastly K being silent in its graduated and transposed written format, was a form of distinctions implemented by English language scholars a few thousands of years later.

I would call the creation of the word Knife therefore something that would not exist in Birtur or Birtur's later developped sister language Turkish, of any of other Turkic family of linguisitcs. In a sense that knife in Turkic would be spelt nife, and would be simpler in sounds, syllable formats and pronounciation. Knife today in Turkish is prounonced with the simpler phonetic sounds (bi-chak), backed by facially derived phonetics, versus more complex words vocally derived.

In any of the Turkic family of languages, even at present times, everything said is written exactly in the format the mouth has prouded it. A-Ta in Birtur, transposed itself to Ata in Turkic and is written exactly as its prounciation. Same with all other words associated with the Turkic peoples language su (water), ka-nat (wing), or el-ma (apple) . Here is a fun exercise, have two different sets of people whose mother tongue is neither Turkish nor English learn both sets of alpahbets, then let them read lets say 50 Turkish flash cards of various words followed by English flash cards. The evidence is there. Turkish will come out beautifully the winner, due to its phonetic simplicity.

Here is another one, in any category of Chinese linguistics which would be any of the languages belonging to the Sino-Tibetan family their alphabet structure requires a minimum of 6 years of formal educational training before children can even start to read, again evidencing complexities humans have done afterwards. In any Turkic school system first graders have already mastered the written art of their language.

Does this mean the family of Turkish languages has no merit in the field of literature, absolutely not what it does mean is that its organizational structure was developped, when no distinctions were implemented. Meaning the language is an easier language to study for its foundation was implemented on natural stages of development versus human interference, altered words and memorization at the more scholarly and complex literary level.

Even more interesting is that Birtur was a language that was not only the first form of communication but it was the most naturally spreading and dominant (lets exclude the spreading of the following languages such as English, French, Spanish due to colonization of the Americas in this equation).

Original Map

Majority are following coastal lines

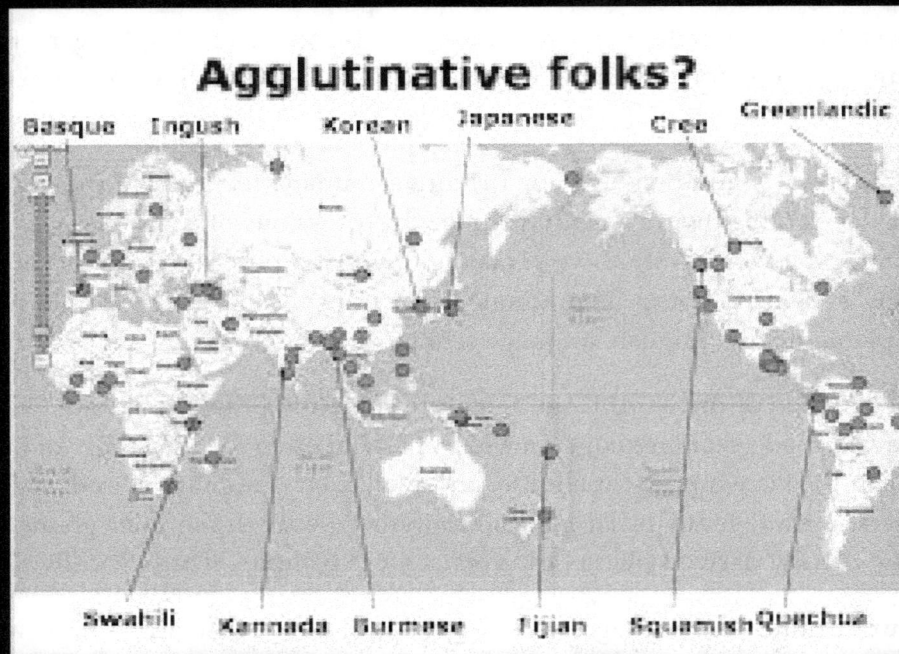

Agglutinative folks?

Basque Ingush Korean Japanese Cree Greenlandic

Swahili Kannada Burmese Fijian Squamish Quechua

(have graphic designer make new based on patterns of Birtur)

Now have a look at the same map but with the addition of pinks spots, which even though some regions are 17,000 years ago came way later and is a perfect examples of what I label overlay. The death of one family of languages based off regions and the start of other langauge.

This occurred due to population increases of clans and the need to implement distinctions to start distinguishing clan members - Lets start from pink point A for example Anatolia, to pink point B Indus valley Pakistan, to pink point C Yellow River China. The oldest regions of the world.

Majority of these travels followed waterways via the ancient silk road with the exception of pink point D the Americas, only because of the collapse of the Bering Strait land bridge. A travel route that went back and forth for centuries. With settlements and their increases, in the name of tribe member identification for example during violent acts, brought forth the implementation of lingusitc individuality and the start of cultural changes in the regions. "I am different, and wish to be different from you ideology".

The forested region of Sahara is newer to that of Gobi which would be our last point. Our goal being to figure out the patterns of our nomadic ancestors, by not only looking at language but the amount of destruction of a landscape compared to the amount of its population.

Agglutinative Languages

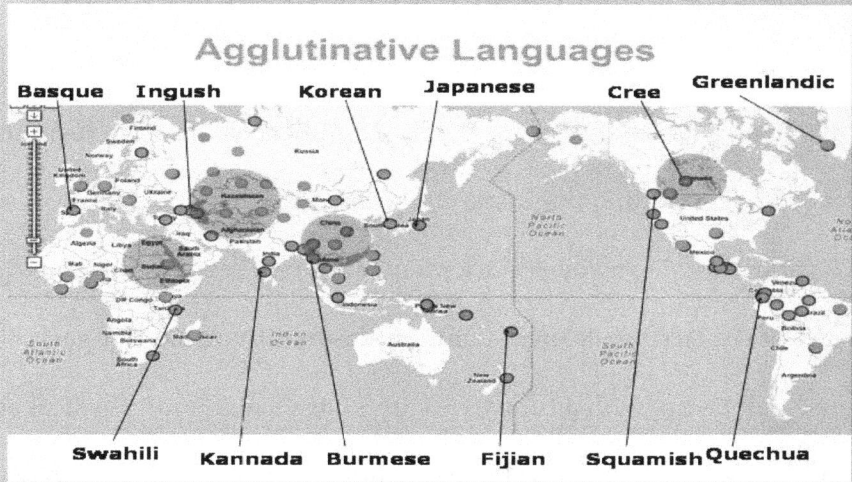

There was an excep
tionThe Americas:

We can trace back the DNA of 6 women only that produce all the native in the americas therefore the americas are an exception because with the collapse of the bering land strait they did not have the population numbers /clan structures as anatolia did to be as destructive as the cradle of civilization was. Again a ripple effect of population growth, continously out of Anatolia. Both regions being equally old .

Clan Growth Comparitive

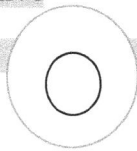

Visualize Anatolia (one small clan)   Visualize Americas (one small clan)

As the world turns and we are also a component of time in our history, change is the only constant in our planetory equation. Furthermore In Anatolia there is a proverb that says "history always tends to repeat itself overtime". Ironically due to advancement in technology and the uniformity of communicated media we are losing 200 languages per year and have many on the endgangered list. The world therefore is reverting back to one form of main stream dialogue

You grew up in a home speaking a certain language for the last 10 years, that place now belongs to you. No one can dispute this. Using this analogy lets envision what 8,000 years does to groups of people living on a land with a certain language. This is called linguistic evolution. The concept that language has now transformed itself into identity and cultural attachments.

Have you not evolved to what you speak is the better rational? If we look at it scientifically we all came from knuckle walkers, but we don't call ourselves apes. We all originated out of Africa but we don't go around saying we are black. As radical as this may seem this is the critical component of understanding linguistic evolution.

Lets pose it differently "Are you a black man or women reading this" now you might be if you belong to any of the African-related communities but if you are not part of this community your first response will be no, with a funny look on your face. If your physical appearance doesn't fall into this category why would your language fall into a cultural attachment for the dialogues belonging to indigenous proto-Turkic.

For many years there is compelling evidence that language may have been hidden to not offend or hurt groups of people. This is a far cry from the realm of what understanding linguistic evolution really means. Especially after a heavy topic like the introduction of Birtur, this section is not designed as a bandage to cover a wound, especially in reference to a language pertaining to what people would associate only to a hotspot country like modern Turkey. The question then becomes what is evolution? Evolution originally referred to as a modification of one species onto a next. Let's take evolution and expand on it now. Let's look at the origins of different DNA's, adaptations to different landscapes, creation of new words to that of a particular environment and lastly understanding time. Combine all these factors and you will get the birth of linguistic evolution.

The Case of the African Shaman Natives to that of the Finnish People:

What if academic professors were to hop on a plane and go to Finland and now tell the locals there, they are now African Natives who speak Turkish. Lets envision the reactions, the Finnish may even get hostile and rightfully so. Why because the Finnish socially have evolved and the errors have been explained below.

1. Linguistically Anatolian Turkish is very Europeanize and should be labelled a *"modified language" after the Ottoman Language*, error number one.

2. Finnish languages contains 2% Indigenous Turkic in it, even though they are linked this is not Turkish, error number two.

3. Many generation later Finnish people have evolved not only DNA wise but also physical appearance and linguistically to their landscapes. This is a distinction and understanding evolution means understanding that we are constantly altering to our surroundings. error number three.

4. If I said something to the Finnish like you are now "African Turkic Natives" they will be offended. the creation of borders created enormous cultural identities. Error number 4

5. The mergence of groups of people on a topography were a form of " investing" of the numerous generations that have come and gone. This is a form of social-patenting in creating their own identity. Error number 5

6. The hundreds of years of modification to their language, and settings created completely new social behaviours. Social behaviours are developmentally innate and are now gone because of lack of proximity to say the Italians or the Asians. Understanding time and the disappearance of nomadic behaviours = new isolated cultural behaviours. Lastly

7.

Let really focus on this because I want to change visual understanding and perceptions.

Versus

Nomadic Turkic language no borders	Settlements social-patenting creation of regions

Awareness when it comes to knowing "the world is not flat anymore", is powerful. The saddest irony in all of this is Turkish itself was completely modified by Ataturk only in 1929.

Now some people may hate knowing their roots stem from a language of a hotspot country or their ancestry is African descent but lets analyze this as well. PUT AFRICAN SHAMAN PICTURE

Firstly what did we say Shaman Turkic and Turkish are related but are very different. Secondly This research is new and whenever large scale progress is pushed towards various social groups backlash is always normal. Someone who has research historical social injustices knows this well. This is why this chapter was done because it will ultimately go hand in hand with National Geography's Genographic project. It is a push towards progress.

My hunch after my research on clans suggest our African ancestry will be highlighted anyways. The question THEN becomes how do we present this via educational progress versus the archaic despotism of patriarchal land ownership and its associated fanaticism. The key response is understanding the birth of what is linguistic evolution.

The objective of this section is to break a perception.

With our ultimate goal in the second addition being: why language was targeted, how our environments fell victim due to lack of awareness or why we should disassociate everything regarding a control of any particular region in the future .

This is called taking the best out of each culture and applying it to present times. In this case it is the incorporation of native beliefs of "no land ownership" to modern times. The futuristic belief of shared ownership and sustainability. Or my very own personal beliefs on what is real advancement which is the complete removal of all international borders. Meaning real freedom and mobility of humans.

For humour let's keep the Olympics though I love the entertainment.

By thinking futuristic at the present moment, if we have truly advanced as humans.

More so discussed in series 2 our goal as humans are to move away from the concepts of ownership and borders, to understand what our environments gave us, why history was modified, to be sustainable, what we are doing and how to advance as people. We will look at what it means to linguistically evolve as a cuture.

According to Smithsonian the development of language says Dr.oshannessy say was a two step process. From parents using three languages in babytalk to children adding radical inovations to the syntax to develop its own language.

radical such a short period of tiime...blah

What changed us from one language to the thousands of "Global Languages" we have today? Linguistic historians indicate the earliest accounts of written documents are 5000 years BC which are Cuneiform writings belonging to the Sumerians. This section will be dedicated to the evolution of human linguistics. The progression of linguistics therefore is defined as starting from the unwritten language of our nomadic indigenous ancestors to its final point today, " *the absolute radicalisation of language*".

The Transfer of Language

Visual diagram of early human surroundings created a human marking or regional label to an area but how did this transferred over to the various categories of early humans that have existed?

We now know the first bacteria of life self duplicated , that every living thing today developed from a form of interbreeding. When different groups of hominids became mutated their language was transferred over as they went through natures, natural selection process.

Before any ships or other forms of transportation, early humans continuously roamed. Now that we have been introduce to Birtur, and the category of language family it belongs to it would explain Agglutinative's dominance as the most naturally spreading family of languages.

Agglutinative folks?

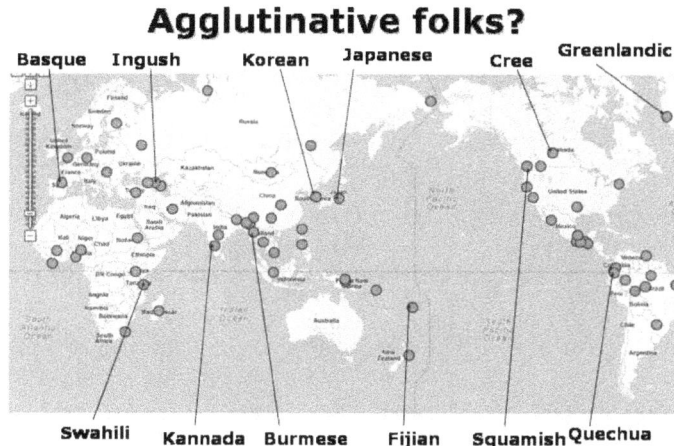

A good example is the Neanderthals which lived in an area having Birtur descriptives, and proto Turkic artefacts. Siberia - translated as by the water's edge si -ileria, and Denisova by lows lands near water Deniz-ova are just some Birtur words pertaining to these regions. Neanderthals and Denisovans inclusive are examples of the transfer of language in different types of early humans.

Geographically starting from the Russian city of Kazan to the Pacific ocean - are communities of Turkic clans which are all unique. These are indigenous communities all culturally, religiously, dialects wise different but what's more remarkable is that they did not go through the process of linguistic overlay or drastic changes to language, see Nomadic patterns why. What is important is that this linguistic spread and their consistency in language evidences one more clue of early humans dialogue during the stagnant period of time.

What is Linguistic Overlay?

- Human Altered Linguistic Overlay - Intentional changes to a language by scholars

- Natural Linguistic Overlay - New topography generates new words, entry into the Americas

Human Altered Linguistic Overlay - The simplest definition is when one language dies out and another one gets brought in. There were several reasons historically for this change. Change in administrations of empires, self preservation or the need to identify during war, colonisations, movement of people due to slavery, and today the uniformity technology brings. In 2014 for example we lose on average several languages in a single year(1), we also have rare cases like the ones in a remote area in Australia where a new language was born(2).

This type of Overlay is the hypothesis that groups of people for whatever reason communicated amongst themselves and commenced altering Birtur era's linguistic dominance. The outcome of change in settlement areas is actually way newer in switching over to other languages, than those isolated in the Americas dating back to the first entry at 120,000 years. These types of alterations are what created a polyphony of proto- type global languages and dialects we have today. An example would be the small pockets of proto-Turkic left over in the Sumerian, Akkadian and Ottoman languages that have existed in the Anatolian Region.

With each new discovery a rewrite has to be done of our history. Linguistic books suggest that everything out of Mesopotamia, pre-Sumerian era belongs to a linguistic label of Babel. This was categorized because of its unknown nature. Today I have suggested historically Babylon's linguistic label Babel be more of a **"transformation period"** versus an actual linguistic. Why because through research I am attempting to evidence the unknown, or what may have been politically hidden. That in Babel's era it is better defined as the transition from nomadic to more permanent settlements; and doesn't reflect the real language of our early ancestors which was in majority an earlier *"classification category"* of proto Turkic, called Birtur.

Lets look at one example of the alteration of language over time. I call this the before and after modification of a word that thousands of years of settlements has changed. Three versions will be presented Diagram 1) one belongings to a nomadic group of our earlier ancestors 2) is an Anatolian based word and 3) is what Europe picked up on route from Anatolia. 4) Cant be found in Mandarin or Cantonese.

Indigenous Wan or Wim is area ⟹ Van in Farsi means settlement ⟹ Von in Europe means higher ranking Baron and their regional controls.   China completely modified and removed the word.

The key here is we are actually working backwards, from the Americas into Central Asia, Anatolia, Africa and then chronologically attempting to order our linguistic history into a timeline. Assessing words, studying complexities of a language, investigating who had alphabet and who picked up other groups of people's alphabets, and lastly the study of phonetics will be our measure of language word order.   This research is completely aided by Harvard University's Migration patterns.

 Now visualize, the changes or words, cultural structures and many other areas over time. Change occurred because of the need to be distinct from other groups of people. This will be discussed in more detail on the chapter on self preservation and why it was so totally transform, from its million years .

Diagram 1   **Diagram 2**   **Diagram 3**

I love the concept of knitting, what is knitting...knitting is taking an idea (yarn) and the goal is the end result (product). Ok what have we learned so far?

The hypothesis

⇨ We came from one point of origin.

⇨ We left Africa and inter-bred and multiplied to form larger clans.

⇨ We transferred language from animals to hominids to humans incrementally over time,

⇨ The transfer of language was initally constant and was incrementally built .

⇨ Scientist indicated are population for a stagnant period of time (over a million years) was around 55,000 people.

⇨ Approximately 500 groups, of roughly 100 people each per group, travelled the globe and settled in small pockets of tributaries along coastal lines.

⇨ What is a million years? United States is 238 years old, a million years therefore would equal 4000 times that of the age of United States, and over 7000 times to that the age of Canada.

⇨ Controlled fires and tools started the standardization of single syllable words associated with what we saw in our environments, call Pictionary linguistics - now known as the language Classification of BIRTUR. These regional names are all descriptive and phonetically can be decoded to the Turkic family of languages.

⇨ Travel time was nothing for our early hominids who for example travelled from Europe to China in exactly 2 years.

⇨ We crossed over to the Americas 120,000 years ago and started creating completely different languages due to isolation and new landscapes, this process was called Overlay. More specifically Natural Linguistic Overlay.

⇨ To understand Overlay better simply divide the planet in 2 - where Africa, Europe and Asia was one land mass while the Americas was another.

⇨ The first homo sapiens, were relatively uniform in physical features; over thousands of years we reached a saturation point with other early humans.

⇨ Settlements lead to population growth. This growth made us intentionally alter language for identification purposes during attacks. A form of self preservation. This is called Human Altered Linguistic Overlay. See Derin Kuyu page xxx and visualize.

⇨ Evolution of language means you have created a new identity over time to the region you have attached yourself to.

⇨ Fortunately for humanity, and so busy implementing distinctions regionally we forgot to change our key piece of evidence "*The Mountains*". We call this the spirit of Chi, their voices echoed in the wind.

⇨ The relative comfort of settlements gave local scholars time to generate alphabets and more complex languages that required heavier memorization or altered words.

⇨ Now we have come to the 21st century - our next topic is the radicalization of language? Can you see at this point, half are sweater completed? We are evolving.

Radicalization of Language

Language has always brought out cultural fanaticism. Are you mad concept, or angry? Do you want to rip up this book because of what is written in this chapter regarding language? If you said yes then this is what's called the radicalization of language. The inability to dissociate a language today to the incremental developments of early humans.

I am a fierce believer in a world without borders, sounds too drastic? It goes back to our need to control our environments down to a square 8x11 cemetery plot. Concepts that never existed with the Native Americans, whose beliefs were based on this land can be shared with everyone. What made the Natives different in thinking than those in central Asia or Anatolia? Simple, something so basic as Overlay.

One form of overlay happened naturally with a change in environment, crossing into the Americas is classified as a new environment. This type of new word generation was brought to the forefront, as our indigenous saw new surroundings and every nick and detail that mother nature created in a topographically different area. An example is seeing an animal that is local only to Canada like a Canadian Elk for example generated a new word.

The other overlay occurred with the intent to self preserve and population growths in settlement. Alterations in language, created engrained radical behaviours in humans. Remembering that advancement can only occur with awareness. Therefore to evolve as humans we need to understand radicalism and root source of this evil.

See Self Preservation chapter xxx for a whole list of how radical we have been in our fight to survive.

Natural Linguistic Overlay- Comes in one format only which is geographical isolation to that of time.

Meaning new landscapes, generated new words.

Example Dodo bird only found in Madagascar- which would have generated a new word.

Over time this isolation started making primitive humans gradually change naturally their dialects and lexicons to their new topographies and create other Pictionary visuals of what they historically saw.

Landscape changes therefore created new words

Radicalization of language

What if everyone spoke the same language?

Imagine each yellow dandelion is one of the Birtur era's speaking 55,000 nomadic people that roamed the earth for millions of year, each having similar language with modified dialects. Lets ignore skin color, physical features and any other types of perceptions that came afterwards, evolution wise there was clearly a need to implement distinctions. Keeping in mine physical features went through its own evolutionary process.

Image clan growths and the need to self preserve, these ever expanding clans. Visualize a bird accidentally dropping seeds and other forms of linguistics starting, the commencement of the growth of trees. This is defined as overlay and as these trees grew it overshadowed anything under it .

A visual example of variant language family creation:

(Diagram 1)   (Diagram 2)

Dandelions are Birtur

Overlay, commencement of other languages and the death of Birtur

(ii) Birtur belongs to the agglutinative family of linguistics. Agglutinative was the **naturally developing and spreading** (gradually changing while keeping its structure) dominant family of languages globally. This is our key clue in understanding some of the patterns of how this dominance was derived due to a stable population which created a relatively stable linguistics.

Birtur's dominance came from the thousands of years of nomadic movement. Always remembering the earth during Birtur's initial era had only approximately 55,000 people. Let's do a comparative between the stagnant population of Birtur for the thousands of years with that of one country the United States. Lets have a look at the following

Global	United States
Language category: Birtur (proto-Turkic based)	*Language category: English*
Logically and naturally developing Pictionary linguistics - vocal responses to visual surroundings	Intentional altered linguistics designed by literary scholars
Learning capacity of later developed (same family) Turkic written- Easier, by 1st grade children have mastered written	Learning capacity of English - Difficult by grades 3 or 4 children still struggle with pronunciation to written capabilities. This is normal in "scholastically altered linguistics".

	** Sino Tibetan family of linguistics is the same. Mandarin, Cantonese etc.
Family Agglutinative	Family Indo European
Birtur era, in majority comprises 2 or 4 letter words here are some examples su, ata, dan, ana, kara, chi, dam, ara etc. * Interestingly facet, little children in turkey suck in air asking for water su. Lets understand how something simple like su (water) started with primitive humans sucking in air, a visual response to a need, turned into the closest representation or vocal to the particular response. Another example is the word similar to boug children will also do that and it is closest to bug (the start of bo-cek) Our first vocabulary started with clicking, mimicking animals, sucking in air and other visual to vocally associated noises.	The other comparative: The natural structure of the English language and with intentional distinction created a more complex innate structure, something common to the indo European languages as a whole. In which one characteristic is the gluing of these Birtur words while applying distinctions and creating more complicated and transformed words. Here is one example: Water (based on the documentary I saw) originated out of a word wa-at-tar-ma. I believe this was 1) intentionally changed from su or si, 2) it was a whole sentence repeated continuously to form the word water today 3) it was glued together to create complexity, most probably intentionally with the realization it was done. 4) This innate grammatical structure is still an intricate part of the English language. Is evidence further by

	one trip to the Mesopotamia exhibit or many history books and how many years ago the British for example glued birtur words.
	Not realizing the words broken up phonetically will give you the patterns presented throughout this book on nomadic dialogue.
Naturally spreading due to Nomadic movement	Spreading occurred due to Colonization
Approximately Population numbers were 55,000 with a larger square footage, of area coverage	Population over 300 Million today has a smaller area of coverage which is normal with our exponential population growths.
Existence Period, 3 million years prior to approximately 10,000 B.C (except what transposed to Anatolia and Central Asia).	Existence Period; When Europe found out of the Americas Columbus 1492
First written transformation format to Turkic came out of Korea	One written format came in Cuneiform writing in Bogaz Koy Turkey.
Extinct, by linguistic overlay, in most part of the world including the Americas anywhere	English flourishing due to technology - creating a uniformity across the Americas.

from 5,000- 75,000 BC years prior. The dates vary based on region. ** Birtur today is transposed in majority to the Turkic peoples language in different dialect formats of Central Asia and Anatolia.	
Contributions to all languages globally- was a starting foundation. Here are some examples: • Arabic some of the basic Birtur words got switch over and includes articles, Al and El for example. • Hebrew first transformation of alphabet was Birtur with Hebrew letters. Also some words in the transferring process are evident in Akkadian. • French, similarity and transformation of certain words example the "pa" in *pa-puk* shoe became pas (foot) in French. Same meaning.	Contributing to other languages today because of media and technology.

- English similarity and direct transfer influence of -From Birtur to Turkic to English words example the words Kent, Basin, ...

- Farsi direct transfer or similarity from Birtur words example Wan to Van settlement and .....

- Chinese from Birtur to Turkic - research needed of the 100 or more Turkic pyramids hidden to the world that will evidence it.

- Native Americas complex linguistics of approximately 900 words were brought in 25,000 years ago in which 100 ck are still critical and in use today.

** With this research and the highlighting of native complex linguistic, the Americas age should be highlighted at 25,000 years old, not the more recent figure of a few hundred years.

We can't already discover very linguistically advanced, habited lands.

Presently based on Oxford write-up over 170,000 words are in the English language.

I believe it would be cultural genocide to dismiss our indigenous history of the Americas. Understanding those 900 words were probably way more during crossing over and also experienced overlay due to geographical distribution. This lead to the birth of our numerous native linguistics that we have today over the remaining thousands of years with its linguistic extinction.	
Comprised of approximately 900 words- based on North American research only	Comprises of over 50,000
Indigenous people with "*Nomadic Tendency*"	Settlement and urban based

After looking at this table now let's provide a good analogy, a southerner's accent in the Alabama and their terminology won't match the accent of that of a New Yorker and their own socially derived terminology or lexicons. No different than the following that the nomadic clan belonging to Jane in Africa won't be like nomadic clan of John in Russia in regards to dialect but the similarities are profound enough to create thousands of years of dominance and regional markings.

There is a secondary critical factor that will back my evidence, we know there existed other humanoid types on earth, and a saturation point was reached, how then did Birtur establish further dominance linguistically over this stagnant period of life.

Language went hand in hand with the more dominant species that came out of Africa. Ever watch any type of Empire documentary or movie? During acts of war when men are killed, women and sometimes children are incorporated into the new empires. Well this also happens in the animal kingdom the other challenging male and his offspring are killed and immediately the winners mate with the females and create their own offspring. What did scientist, we have a shared ancestry.

Maps below represents a stable period of Nomadic movements for the 23-50 clans that roamed earth and their regional territories.

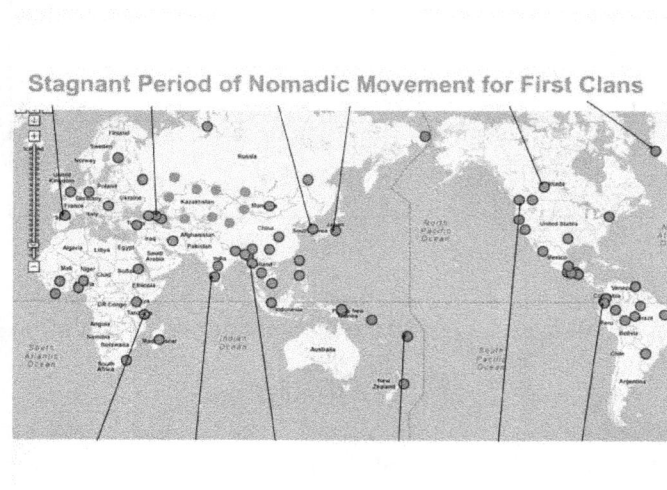

Stagnant Period of Nomadic Movement for First Clans

Keeping in mind that one man was able to travel from Europe to Kamchatka peninsula in 2 years and humans have been roaming earth for 3.2 million years, we as humans have to acknowledge this world was very small for our nomadic travellers.

Changes that occur

Overlay- Major historical changes

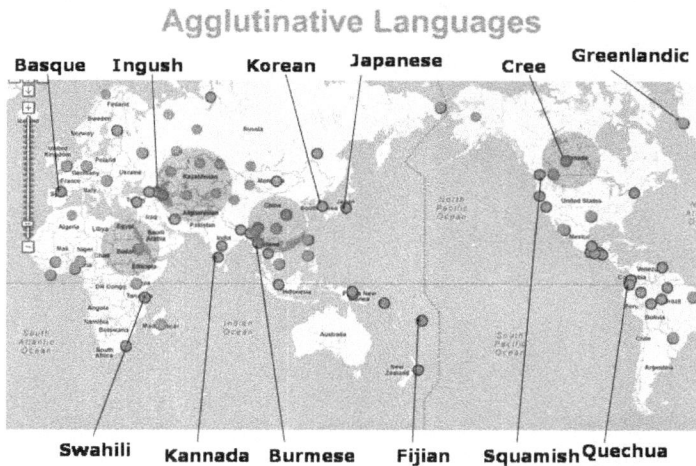

Agglutinative Languages

Basque Ingush Korean Japanese Cree Greenlandic

Swahili Kannada Burmese Fijian Squamish Quechua

Proto-Indo European was also Agglutinative and assist us in understanding this overlay

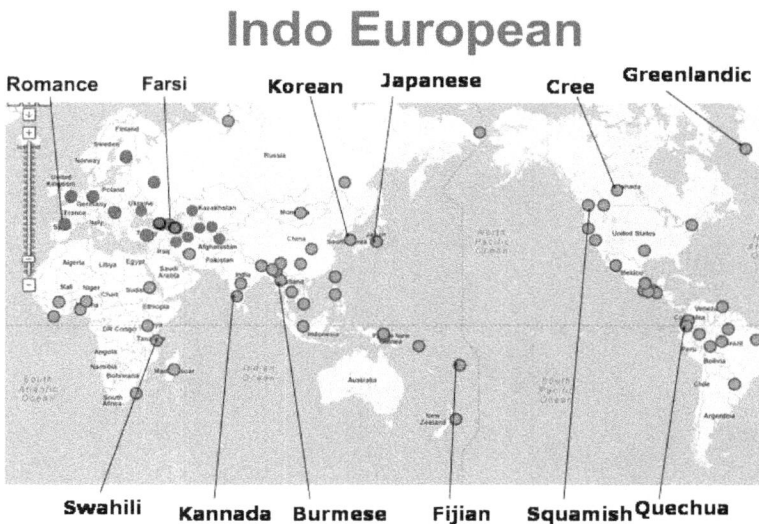

Indo European

Romance Farsi Korean Japanese Cree Greenlandic

Swahili Kannada Burmese Fijian Squamish Quechua

Today the agglutinative family with its different varying dialects but the language which has the

same format for grammatical linguistic makeup.

Agglutinative folks?

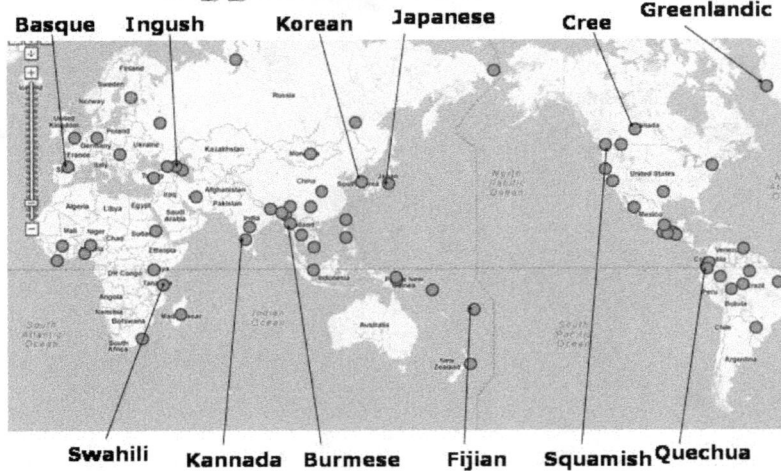

Basque Ingush Korean Japanese Cree Greenlandic

Swahili Kannada Burmese Fijian Squamish Quechua

To

(iii) The definition of overlay continues

	Overlay of Sumerian, Akkadian, Elamite	
	Overlay of more Modern Linguistic - also known as the building of other languages on top of Birtur.	
The word al	The word "Al" in Arabic is an article however it probably means an action "to give" in proto- Turkic, most likely the action of giving everything to the shamanistic	Al was the start of double syllable linguistics for Birtur nomads that got transposed linguistically to an article in Arabic.

deities which still has a resemblance to customary traditions in Anatolia. Everything done in the name of God.	Arabic formatted the two words and then merged it together as an article. Al-mahara AL-jumhriyet AL-Hamidiyya AL-Dayaa
Same with the word el which is another Arabic article but means "outside of clan / foreign" in proto-Turkic once broken down. These words now articles were transposed from proto-Turkic to Arabic. Today el also means hand.	El is the start of double syllable linguistics in Birtur which got transposed to an article in Arabic. El-kumbra El- kheima

(iii) We don't have any point of origin for the linguistic of these nomads because they were a moving group of first humans. This proto-Turkic linguistic evolved with the interaction of the first clans themselves. However we can deduce the following:

(i) From the time the first humans reach 2, 10 or 50 in numbers	Some format of dialogue commenced.

(ii) Oldest fossil found	That would be Ethiopia
(iii) Pictionary words of some regions where that fossil was found	Sudan - aka made from water in proto-turkic

which that Pictionary would be accurate since that area was all marshy and wet. Water tables found a little north in the Sahara. |
| (iv) Concept of time | The need to think differently when we look at human linguistics. Historians did not have artefacts to give a more accurate depiction that another type of linguistics existed. But a comparative of some indigenous people in South America simply move around with one basket only of tools but they have a relatively strong linguistic dialogues. |
| (v) The evolution of skin color | Due to centuries of colonization, wars, violence, implemented borders, nationalism and distinctions made, the human-mind has naturally colour-coded people which made historians, not notice the patterns of this proto-Turkic language that I labelled "Birtur".

A mental picture of the larger linguistic jigsaw puzzle was difficult to piece together because of our modern perceptions. The notion that different shades of clans of people (that was relatively stable) for millions of years |

| | couldn't be possible with the concept of them having one linguistic only.

A group of people that only left behind these invisible marks of their existence with the "**consistent patterns**" of their regional names globally, and what they saw millions of years ago.

I believe it is a critical component of human history that was missed. In addition and this is where the environmental portion comes in, our environment that these nomads, marked by labelling their settings, had natural check and balances to keep this population of people relatively stables for millions of years.

My analogy to that, of the "Amazon Jungle case-study" below will elaborate how dangerous human intervention to save life is and why reproductive control is needed in our environments today. (INCLUDE DESTROYING BIODERSITY) THE COMPASSION WITHIN US TO PRESERVE LIFE DESTROYED ALL ELSE. |
|---|---|

If we still want to act primitive lets scrap language altogether and continue doing ululations...

(iv) The large scale geographical distribution from Korea to Italy, Siberia to North Africa, North and South America provides evidence how nomadic our ancestors were and they would have had to have a stable linguistic amongst themselves to marker areas for possibly millions of years.

So the question is what changed? Let's remember the answer, which will be elaborated further below: Settlements, population growth and the need for human distinctions, for self preservation.

** 6 women produced all natives north and south, 1 blue eyed women produced all blue eyes

note keep it out of politics

Now let's look at these markers.

The world is small for those who chase herds. Remembering that in one lifetime clans can travel from one end of Europe to the other end of China at least 3 times or more. Piecing together through Pictionary visuals the only evidence of what our nomadic ancestors left us. Here are examples of some regional markers in their most primitive format. Please keep in mind because of the geographical distribution and varying dialects, I am only able to pull out patterns that are consistent based on the importance of the words.

Please keep in mind I am an environmental researcher who is looking at the notion of how distinctions are associated with population growth and its impacts on our environment. These are my personal findings and it may take many years to research each region from a historical perspective. Therefore I am publishing only some of the patterns that I see, and that there existed a proto-turkic group of nomads before an overlay occurred of other global languages. My hypothesis is as follows that population growth contributed to some of our most destructive behaviours in history including the change in linguistics in clans. Lets learn by looking pass the dialects in the lists.

** keeping in mind some of these words were modified by the ottomans as civilizations advanced but I still incorporated them as a transitional - evidencing the complex linguistics of Palaeolithic people as they roamed.

Fortunately even though humans have felt the need to apply distinctions, and we have changed most of the original historic names to the languages that have come afterwards we are able to analyze these patterns because the mountain names in majority have stay the same. My hypothesis being the mountains by chances were never untouched.

Key words that are consistent globally

Ata (leader), Kan (blood), Kus (bird), Kanat (wing), Suffix Van / Stan (settlement), Su or Suyu or Si (water), Kum (sand), Basin pronounced ba-sin (basin), kaya (rock)

Now lets look at how dominant proto-turkic from the agglutinative languages were. What do you notice funny of this map?

The rule of thumb is let's look for geographical distribution, dominance and patterns, and why did those linguistic patterns changed. What would prompt large countries like china for example in the 15th century to destroy maps

Put map 1-2

Agglutinative folks?

Basque   Ingush   Korean   Japanese   Cree   Greenlandic

Swahili   Kannada   Burmese   Fijian   Squamish   Quechua

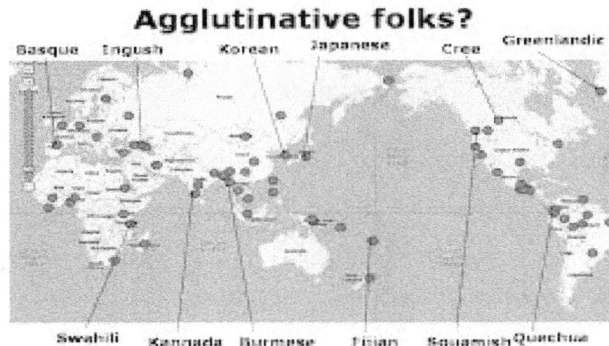

EXPLAIN RESEARCH IN COMPARITIVE WITH ENGLISH FRENCH AND MANDARIN why it occurred -Why these are way more difficult

explain overlay concept what happened.

Dandelions on a flat field - a bird drops two or three maple tree seed that eventually grows and overshadows the flowers- the branches are global linguistics.

Each branch now representing a language. My hypothesis of why this happened firstly geographical disconnect and secondly population explosion and scarcity commenced. Protection of immediate clans were needed.

example research

Recently it was discovered...English was born in the cradle of civilization once that settlement hit 50,000-60,000 in numbers.....explain what happen and the desperation for self preservation. lets visualize how difficult it was without machinery, without equipment or anything to maintain a habitat of that many people in a region of the world that was continuously under attack. The nomadic patterns of our ancestors and why the south moved north. and why today in that same cradle there is a town that was built 13 stories underneath the earth. Derin KUYU

.

Proto-Turkic is very simple and logical in order and is still the case with modern Turkish today. Lets continue with the markers, what do you else do you notice that some words are more dominant.

animals communicate too

first words

start of evolution of religion from human sacrifices we have come so far lets push that up a little

PICTIONARY PERIOD

first words zara (sara)

zara-ustra India, zaragoz Spain, (ne zarar?- transformed to nezar middle east),

ata, anu, kayuk, papuk, kus, cus, gus, kaya, su or si, kan, sara (zara) today zarar, cam, kuyu, kok, hua, gua, dua, gok,dan, can, God = kangri tengri tan, tapana

put native marker picture, branches,

the words from millions of years before that nomads of earth left behind, a linguistic that was thought of as being too basic or animal noises because no written. consistent visual markers are the only things left.

before india there was human sacrifice called zara-ustra the one who gives zarar by knife ustra

gobi and sahara might have destroyed everything prior.

did humans underestimate the timeline of human communication, what is 5000 years in a pool of millions of years?

Argument

my hypothesis. indo occurred after buyu in korea and black sea separating massive famine.

why not indo european like previously assumed

geographical distribution of agglutinative linguistic

too wide

sumer, elamite, akkadian all have underlay of proto-turkic

BIGGEST EVIDENCE PROTO INDO EUROPEAN IS ALSO AGGLUTINATTIVE

doors of babi-illim babylone have proto-turkic

siberia 40,000 before christ = means turkic authors evidencing this.

back up critical- 2 professors works and 1 author siberia

mexico article of turkic words found find it

native words = 900 prototurkic work

i.

ii. why chinese and english and other european languages are so difficult- the design was done intentionally

iii. how the transfer came about of the groups of languages diagram

iv. our linguistic history and language progression

v. let me provide the patterns , the phonetic its breakdown, and evidence to you.

vi. sumer - translate the phonetics

vii. languages of turkic- regions of the world:

viii. The world was small for people who followed the herds, i call it turkic for the ease of this publication. first large clan roam, can be found egypt, spain italy, india russia, saudi arabia, china..list Korea, civilizations, Africa, native 800, give 10 examples Pictionary locations. once the patterns are shown it is easier to follow our ancestors.

ix.

x.   why it may have been hidden,

xi.   the start of pictionary

linguistic families......the order

1) the spread of agglutinative family of linguistics is the dominant one lets understand what dominant means. The one that spread out the most. The impacts of violence, warefare, sexual disease and the birth of religons.

2) Start of semitic.

3) Proto indo European is also agglutinative 15,000 years ago.

4) indo european

3 big regions of historical changes in linguistics middle east(semitic), china, Anatolia(indo european)

warfare hardship for our ancestors, the psychology of primitive humans,

•

• Single words, can you see the step-up from ululations. Lets imagine our ancestors now as they grew, became a clan and roamed the planet, let's try to imagine what would have been critical to them in finding each other or calling each other. The answer is not complex here, it would have been Pictionary type words. Words associated with locations or words associated with the hierarchy of the group.

Book states out of all the indigenous in the americas they are made from a handful of women, 6 from another article I had seen. they kept transferring over, these were indigenous peoples regions.

remember mamooth

Markers used to signify location of settlements or prayer, usually found near mountains. The goal is working backwards on the migration map, using native history to see what life was like before overlay.

Location the same, structures have changed, in other parts hundreds of structures destroyed because of religious conquests. Unfortunately the most recent 2012 Buda temple in Afganistan.

Silk road farsi.

we can deduce our nomadic ancestors would have made up, a group anywhere from 50-100 clan members that continuously roamed the earth for food.

It is the history of their nomadic patters, communication and unwritten linguistic. What they visually saw, the dynamics of their basic language, how some words where consistent, and the start of temporary then permanent settlements that led to overlay of other languages and who we are today.

Linguistic Evolution theory acyronms case study

the concept that in 200 years radically changes occurred. we are not talking ethics and the movement of people from slavery.

Case Study of Polynesia locals versus. The Caribbean no intermixing

rabbits 2 one generation

african american physical changes to landscapes 200 year very minor

Radicalization of language

his is the main thesis of this research, from an environmental perspective how do we all become sustainable in a more advanced format without primitive behaviours such as violence or islands of trash developing? How do we teach the masses our one point of origin connection and the need to pool together to resolve a bigger enemy that this one super-organism called earth needs medicine to battle our destructive behaviours? My one generation fix theory will be highlighted further down which goes hand in hand with this thesis.

Why I am deviating from linguistics and talking about environmental matters because language is associated with cultural nationalism and sometimes fanatic social behaviours and I wanted to make sure this little section was inserted so the publication creates awareness on its clear objective and doesn't fluster humans with anger.

The radicalisation of language is associated with domination control of land down to a cemetery when the world was not at 7 billion

----

Why I am deviating from linguistics and talking about environmental matters because language is associated with cultural nationalism and sometimes fanatic social behaviours and I wanted to make sure this little section was inserted so the publication creates awareness on its clear objective and doesn't fluster humans with anger.

In saying that let me provide three examples of the transposing of Birtur words onto other linguistics and there are many, the first is the word Basin. In English it is defined as a geographical description of a land drained by a river and its tributaries, today however in modern Turkish it means publication but most probably meant lower step by its etymology origins of the Altai Mountain regions.

The second is the Birtur word Papuk, it got transposed to the following a mountainous region in Croatia, but refers to hand sown shoes all across Anatolia and Central Asia and can be broken down to the french word pas - meaning feet, by using the first syllable of the Birtur word Pa-puk.

•

2 word

Egypt tutan amen give examples holding god in sky

Korea chang dok palace - can dok - spill life

Korea kyongbok region koyun bak look sheep area

full sentences Mesopotamia British ne bu cha nezzar - what is this striking evil eye ref lightening

1 global singular language when population was 1 million

several languages with population increases, stresses, need for distinction in ware fare.

Bibliography

7-

8- Evidencing other researchers work list all the authors siberia, korea, italy, mexico, north america find them

Index

Progression of Languages

Birtur is the foundation of all languages, a form of proto-Turkic that transposed itself from 1 to the 6,909 languages we have on earth today. In other words what the Greeks gave to the Europeans, is what Birtur gave in terms of the first vocals and syllables, to the global languages we have today.

How can I be so sure? When we look at linguistics we should also look at its evolutionary stages, and its complexity levels.

Birtur words started off as Pictionary Descriptive for one thing. Secondly they are logical in order regarding lexicom structure, this was done when the world had a stagnant population and no real need for the implementation of distinctions. I would be referencing linguisitc distinctions in this particular case.

A language that is agglutinative in structure and is vey simple phonetically references the following, that what ever is sounded out, eventually became exactly its equivalent in writen format. When we look at English for example its complexities evidence highly developed "human altered" language modifications to its predecessor the agglutinative linguisitc structures.

Advanced gramatical formats such as those pretaining to the any of the Indo-European family of languages (English, German, Farsi, Greek, French, Armenian) confirms that these language belonged to grammatical structures that came way after. It would also clarify why all Proto Indo-European languages were once agglutinative in structure and one of Semitic's proto format for example, Akkadian, also has many Birtur terminology in its language. A key in understanding the order of human language development.

Take for example the word knife, very misleading for those attempting to learn English which should be prounounced or spelt like nife. In Birtur this word would have never existed for three reasons.

The first being it is not simple enough for primitive humans to produce (not in the language developmental stages which were facially produced expressions of first noises, to that of single syllables), secondly the word knife if we visualize it, is not a descritptive word of that stagnant period and lastly K being silent in its graduated and transposed written format, was a form of distinctions implemented by English language scholars a few thousands of years later.

I would call the creation of the word Knife therefore something that would not exist in Birtur or Birtur's later developped sister language Turkish, of any of other Turkic family of linguisitcs. In a sense that knife in Turkic would be spelt nife, and would be simpler in sounds, syllable formats and pronounciation. Knife today in Turkish is prounonced with the simpler phonetic sounds (bi-chak), backed by facially derived phonetics, versus more complex words vocally derived.

In any of the Turkic family of languages, even at present times, everything said is written exactly in the format the mouth has prouded it. A-Ta in Birtur, transposed itself to Ata in Turkic and is written exactly as its prounciation. Same with all other words associated with the Turkic peoples language su (water), ka-nat (wing), or el-ma (apple) . Here is a fun exercise, have two different sets of people whose mother tongue is neither Turkish nor English learn both sets of alpahbets, then let them read lets say 50 Turkish flash cards of various words followed by English flash cards. The evidence is there. Turkish will come out beautifully the winner, due to its phonetic simplicity.

Here is another one, in any category of Chinese linguistics which would be any of the languages belonging to the Sino-Tibetan family their alphabet structure requires a minimum of 6 years of formal educational training before children can even start to read, again evidencing complexities humans have done afterwards. In any Turkic school system first graders have already mastered the written art of their language.

Does this mean the family of Turkish languages has no merit in the field of literature, absolutely not what it does mean is that its organizational structure was developped, when no distinctions were implemented. Meaning the language is an easier language to study for its foundation was implemented on natural stages of development versus human interference, altered words and memorization at the more scholarly and complex literary level.

Even more interesting is that Birtur was a language that was not only the first form of communication but it was the most naturally spreading and dominant (lets exclude the spreading of the following languages such as English, French, Spanish due to colonization of the Americas in this equation).

Original Map

Majority are following coastal lines

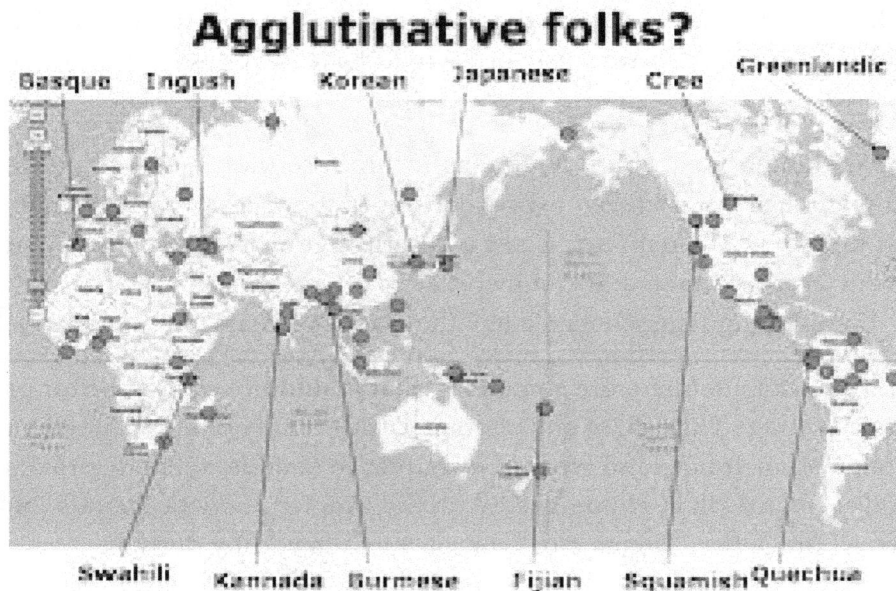

Agglutinative folks?

Basque Ingush Korean Japanese Cree Greenlandic

Swahili Kannada Burmese Fijian Squamish Quechua

(have graphic designer make new based on patterns of Birtur)

Now have a look at the same map but with the addition of pinks spots, which even though some regions are 17,000 years ago came way later and is a perfect examples of what I label overlay. The death of one family of languages based off regions and the start of other langauge.

This occurred due to population increases of clans and the need to implement distinctions to start distinguishing clan members - Lets start from pink point A for example Anatolia, to pink point B Indus valley Pakistan, to pink point C Yellow River China. The oldest regions of the world.

Majority of these travels followed waterways via the ancient silk road with the exception of pink point D the Americas, only because of the collapse of the Bering Strait land bridge. A travel route that went back and forth for centuries. With settlements and their increases, in the name of tribe member identification for example during violent acts, brought forth the implementation of lingusitc individuality and the start of cultural changes in the regions. "I am different, and wish to be different from you ideology".

The forested region of Sahara is newer to that of Gobi which would be our last point. Our goal being to figure out the patterns of our nomadic ancestors, by not only looking at language but the amount of destruction of a landscape compared to the amount of its population.

Agglutinative Languages

Basque   Ingush   Korean   Japanese   Cree   Greenlandic

Swahili   Kannada   Burmese   Fijian   Squamish   Quechua

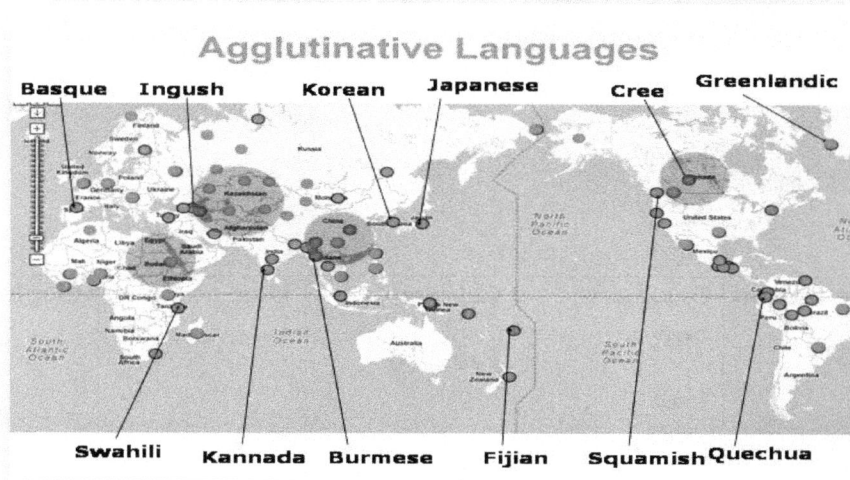

There was an excep tionThe Americas:

We can trace back the DNA of 6 women only that produce all the native in the americas therefore the americas are an exception because with the collapse of the bering land strait they did not have the population numbers /clan structures as anatolia did to be as destructive as the cradle of civilization was. Again a ripple effect of population growth, continously out of Anatolia. Both regions being equally old .

Clan Growth Comparitive

Visualize Anatolia clan)

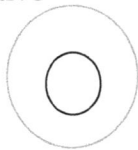

Visualize Americas (one small

As the world turns and we are also a component of time in our history, change is the only constant in our planetory equation. Furthermore In Anatolia there is a proverb that says "history always tends to repeat itself overtime". Ironically due to advancement in technology and the uniformity of communicated media we are losing 200 languages per year and have many on the endgangered list. The world therefore is reverting back to one form of main stream dialogue

We can conclude that the many variations of languages family out there occurred for the same protection of clans as populations started to increase. Intentionally changes

For Fun Game

Create your own language

Apple is now Kun	Banana is now Cun	Cat is now Pao	Dog	Elephant	Fat
Green	House	I is now cik	Joke is now Gae	Kite	Lemon
Me	No	Open	Patato	Queen	Rabbit
Sugar	Tie	Umbrella	Violin	Water	Xray
Yellow	Zebra				

Now before you laugh Australia's children did this and today it is a language. explain

Homo Sapiens	• Basic modification of first sylables to extended words, in more than one language today. Permanent Settlements and general locations of Proto Turkic Urheimet words frequently seen HAN & SHAN (Central Asia), DAN, (Africa) CAN (Anatolia) • Duplicate words transferred to english as well At, HAT, YAT, BAT, CAT
Homo Sapiens	• Followed by two words conjoining, start of Overlay , and variations • Al and El are both beginner proto turkic words, Al Paca (name of food /camel) in Peru • Today transposed linguistically as articles example Al in Arabic / EL to Elle in French language today .
Written	• Sumerian written text mesopotamia
Today	• Full languages and dialects of the world today • from 1 to 6909 total • Changes predominantly occurred in last 15,000 years with population spurts.

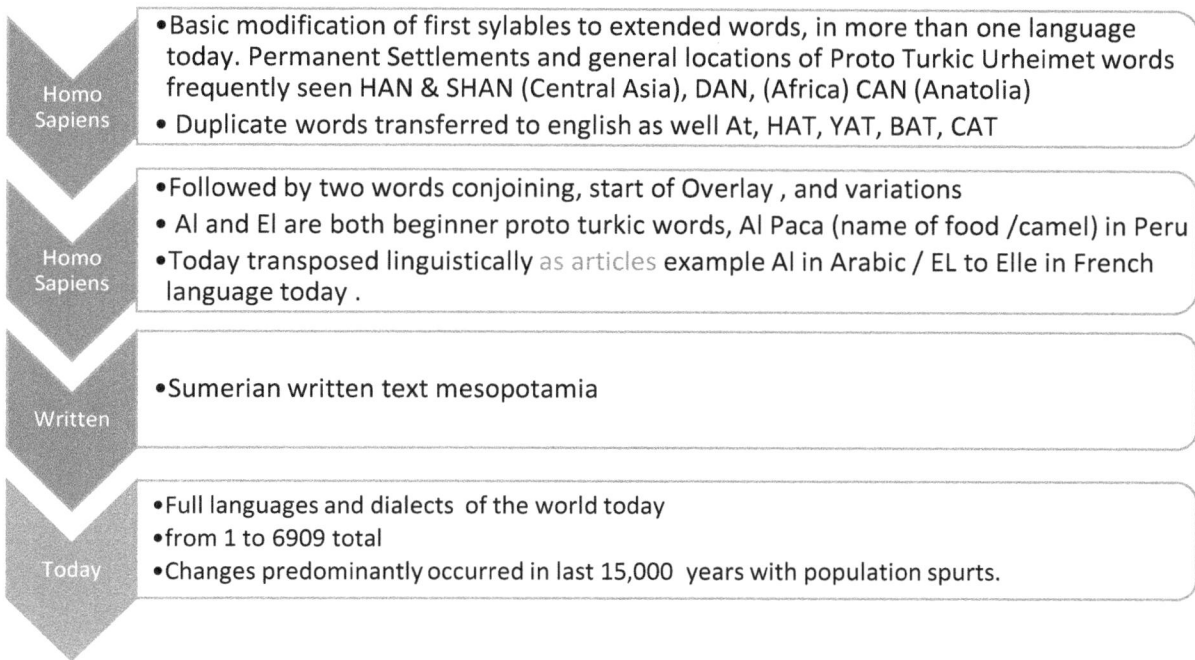

Bibliography

1) national geography

2) SMITHSONIAN INSTITUTE

The Chronology of Language

" *Turk and a meaning similar to that of barbarianism or heathen*" *is famously historical. Well now we know why?*" *It is its age.*

gobekli tepe 15,000 proto-indo european (WAS AGLUTINATIVE PART OF SHAMANIC PROTO-TURKIC) commencement of transfer but could this be wrong? look at a linguistic map below.

southern turkey analogy 20,000 years co-existing give case study why

afrikana development of co share words - indo european 30,0000 years *based on size tree ring of indo european that only cross referencing of DNA and linguistics will merge the patterns out*

iran possibly 35,000 daschet dessert was deschet (deschet is too advance turkic)

a project my estimates would take several years to decode once every human submits their DNA.

islamic verses and farsi (indo) connection to africa

indo European Africa kilamjaro mountains Kiswahili ottoman language and Islam the connection.

are age related and historically ignored in the mind of European linguistics. Who at the time couldn't grasp dark indigenous to actual dialogue. they looked at Anatolia for clues of their earliest history. bypassing the entire landmass. Role of race

today genographic project, clan research etc will change that perception. No different than changing the swastika perception belonging to Hitler when it was an indigenous symbol, stolen by the Natzi's. Give it back we say not just the symbol but the heritage belonging to Africa. biblio

write about ululations link to early indo European. transformation of screeches to rapid thrill

Review

Start
- When the world had one language
- Grunts and Noises in its understandable variations, between animals
- Key animals have full dialogue

Predessors to Humans
- Mimicking other animal noises → Common in the animal kingdom, even amongst early humans.
- Rolling of Tongue, Clicking for hunting purposes.

Hominids
- Intro to the concept of Pictionary Linguistics
- Simple first sylables descripitive word examples Si, Awa, Ova, Ata and the meaning behind ululations
- Basic Conjoined words

Human
- Advanced sentence strucutres
- Overlay and variance of language development
- The 1 to 7000 ratio for languages is equivalent to 55000 to 7 billion populations ratio anlyses

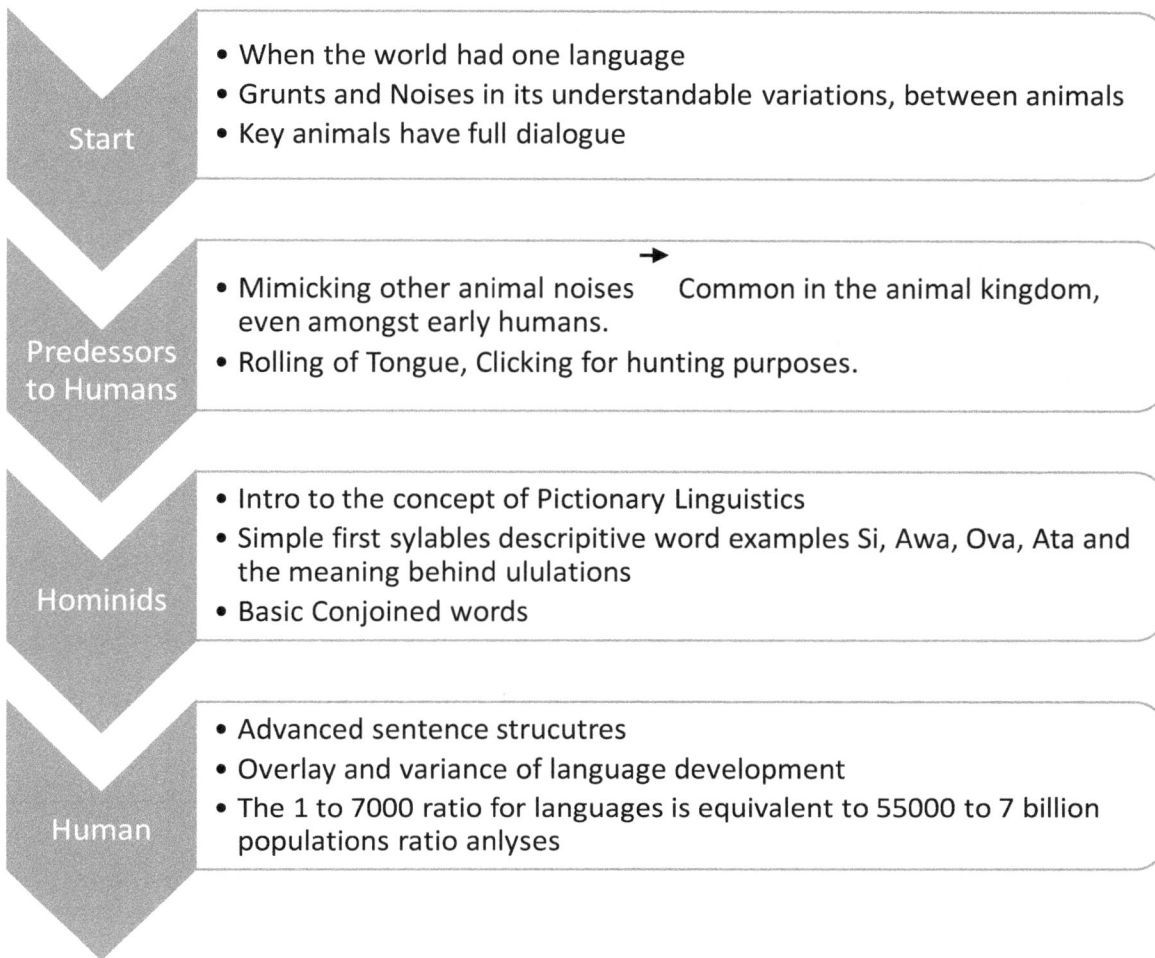

And now let's Chronology of some languages and their associated families.

The chronological age of first clans and their associated language.

indo european in africa is equivalent to its proto format of proto turkic in africa

kis.wah.ili and kilamjaro and nigeraian clans and ottoman emprire. evidence the wrongs of hiding farsi in africa. developped in anatolia english water

- ➢ Agglutinative dominant naturally (working backwards from the Americas)
- ➢ Simplest proto indo European was **agglutinative** - birtur is agglutinative source,
- ➢ Sumer and akkadian have understandable words examples.
- ➢ Ak and Akad are well known proto Turkic first syllables- suffix ian means brothers - there is an overlay process here indo European ending a change is occurring. large

groups of people(brothers) speaking ak or akad (agglutinative mix with indo European source)

➤ the lexemes found in other language pertains to birtur descriptives and first syllables

➤ visualize

make a round circle on cement, throw 100 blue marbles making it dominant. now take a bucket and throw 50 green marbles. overlay

the blue marbles represent the spread, our furthest point on this planet are birtur classified in nature. green is the change but not reaching all points.

Example coldest point a descriptive words Siberia is si-leria by water's edge
The farthest points finland, siberia Kazan to Kamyatka peninsula are all indigenous Turkic tribes.

Educational Information/ Interesting Fact

Anatolian Turks are more European, Turkic are indigenous people, their dialects foods customs religions are different

Notes EXPLAIN BETTER dissect this better
Lets use a mountain as an example. Italy ALBAN.
AL BAN
Take food green leaf wrapped in grains to be exact

so we have agglutinative   Al ban - direct simple command to only 1 person
indo European   english   take ban - indirect who are they referencing
As language advanced humans realized over time that there was verbs, agglutinative are easy. They take a word and add on a suffix. English completely altered the entire sentence.

Al ban istedi	They wanted ban
Istiyorum clearly defined an entire sentence can be one word	To want who?
istedik	We wanted
Istiyince -1 word	When we want - 3 words

we cant do this in the english language for example

- add the mountains
- study populations numbers
- ease of phonetics
-

" Turk and a meaning similar to that of barbarianism or heathen" is famously historical,

Well now we know why?"

The chronological age of first clans and associated language.

➢ Agglutinative dominant naturally (working backwards from the Americas)
➢ Simplest proto indo European was **agglutinative** - birtur is agglutinative source,
➢ Sumer and akkadian have understandable words examples.
➢ Ak and Akad are well known proto Turkic first syllables- suffix ian means brothers - there is an overlay process here indo European ending a change is occurring. large groups of people(brothers) speaking ak or akad (agglutinative mix with indo European source)
➢ the lexemes found in other language pertains to birtur descriptives and first syllables
➢ visualize

make a round circle on cement, throw 100 blue marbles making it dominant.
now take a bucket and throw 50 green marbles. overlay

the blue marbles represent the spread, our furthest point on this planet are birtur classified in nature. green is the change but not reaching all points.

Example coldest point a descriptive words Siberia is si-leria by water's edge
The farthest points finland, siberia Kazan to Kamyatka peninsula are all indigenous Turkic tribes.

Educational Information/ Interesting Fact

Anatolian Turks are more European, Turkic are indigenous people, their dialects foods customs religions are different

Notes EXPLAIN BETTER dissect this better
Lets use a mountain as an example. Italy ALBAN.
AL BAN
Take food green leaf wrapped in grains to be exact

so we have agglutinative   Al ban - direct simple command to only 1 person
indo European   english   take ban - indirect who are they referencing

As language advanced humans realized over time that there was verbs, agglutinative are easy. They take a word and add on a suffix. English completely altered the entire sentence.

Al ban istedi	They wanted ban
Istiyorum clearly defined an entire sentence can be one word	To want who?
istedik	We wanted
Istiyince -1 word	When we want - 3 words

we cant do this in the english language for example

- add the mountains
- study populations numbers
- ease of phonetics
-

Chronological Order

notes

when we look at language ignore population numbers to a degree, because of the concept of overlay and wars. Instead look at the distribution of language. there is a direct correlation between global distribution of a language and its age.

what is 2million years to that of an acceleration in the birth and spread of languages which is 15,000 years to date.

ok we have diverted that is history too, preserving language and culture today

understanding time

why did it happen

the domino effect from the animal kingdom (2 meanings to language animal and human) but it came from animal.

what we have done to the concepts of language

Fanaticism associated with language- if i dont kill you you will kill me, wait a minute that was thousands of years ago beliefs.

why these distinctions of language were so needed 15,000 years ago versus technological advances and why this detrimental ideology has to change.

how language and race and Genographic project can make a better place knowing your ancestors were black, proto turkic, and indigenous for the course of two million years born out of Africa

why studying descriptive words will aide us in preserving our history.

why people with these languages were hated. show evidence

again do you understand time what 15,000 years is? now focus on 2 million years

zambezi death cloth origins out of africa, stolen by islam hidden by christianity, completely ignored by judaisim. (sterling not knowing uka out of ethiopia) word origins of hebrew. ethiopians are not white darling.

we are too advanced now. we are not in mandela's era to turn back.

Human can use the study of linguistics to aide us in understanding history, politics, movement of nomads, and many other areas of research. There are 3 families of linguistics that are the oldest in terms of development born in Africa, Anatolia and Central Asia. These categories are as follows.

➢　　　Agglutinative languages For example Turkic, Japanese, Finnish Languages

➢　　　Semitic languages For example Hebrew and Arabic

➢　　　Indo-European For example English, Greek, Kurdish & Armenian languages

➢ Now let's remember these three major categories of language families which derived from the interactions of the first civilizations of the world.

Criteria

Alphabets

⇨ First component who has alphabet who doesn't we are now putting in order languages.

⇨ 2nd component Time it takes to learn a language (based on alterations). Chinese grade 6 students still have difficulty gasping difficulty in alphabet and written versus Turkish students grade one has master written.

⇨ 3 rd component using the Genographic diagram and working backwards to assess languages

⇨ 4th Looking at language families - which was naturally spreading, which was due to colonization

⇨　5th who had visuals descriptives in their language

⇨ 6th alphabet numbers, lexemes and phonetics

If you categorize the above then you get the following languages in order:

- developmental gap in language between Birtur and proto Turkic birtur is all visual no alphabet south american indigenous

- proto Turkic in <u>certain regions</u> like basic tamgas in Korea / Etruscan / Sumer had alphabet

- proto indo European / no alphabet 15,000 b.c language came from Farsi Persian empire huge. however family structure is agglutinative.

- Farsi / Kurdish no alphabet because of ottoman picked up Arabic 8,000 bc to keep up with neighbouring countries

  countries with alphabet

- Anatolian turkish latin alphabet only born in 1929 just developed alphabet when you have alphabet you are very new on the human timeline

- Akkadian 10,000- had alphabet

- Semitic 6,000 bc ck hebrew had alphabet

- Armenian Greek had their own alphabet

- Byzantium had their own alphabet

- Sino Tibetan 10,000 BC. check most complicated elementary 1-6 to learn to write extremely complex alphabet

- All european countries have their own alphabet

bibliography

1. wikipedia definition

I was conceived in the Ararat Mountains- I study Mountains

The Mountains That Whispered

When the world had no shovels, and their souls left this planet, their families fed the mountains. Today their voices echo in the wind.

A section that is phonetically perfectly translated.

Before I introduce this section I would like to highlight everything I do, absolutely everything is to push global environmental sustainability. This section is dedicated to our indigenous ancestry and a reminder that we should look up to them regarding our harmony with earth.

Understanding that settlements like Summerian as old as they may seem are so new compared to the 1-2 million years our earliest animal proto type humans, have been speaking.

This is during the stagnant period of life, when environments and animal noises contributed to first words, and dialogue. Its final outcome being it transfered over to the many languages we have today.

500 total globally disperse communities between 2 land masses before overlay upto 15,000 BC - Then switch occurred globally. Following overlay and the mountain tributeries here is the list.

** put how arabic articles were born.

ck perpetua, dilgua italian negatives

Starting with the Spiritis of Africa

1. Al ban Italy / Take Food feeding their Gods. Ban is a proto turkic word for shamans: Banarasi is a food in india decoded - food in between ban -ara si

2. Zor zor moutnains Morocco / hard hard mountains

3. Tan Tan Mountains - Tan is a proto abbreivation for tanri god all under agglutinative tan, ban , can , kan family

4. Achim Germany / hungry region mountain names

5. Carthage Morocco -historic name Etruscan Cara-aza / Dark soul

6. kapan armenia/ the one who snatches afterlife connotation

7. zangezur armenia/ zan-gezur not sure what zan is ck most probably modified from tan or kan but gezur is too travel again connotation to afterlife. zan found in africa rhymic correlation to kan, tan and ban.

8. karki azerbajan /bird today means crow in modern turkish but most probably was referencing black vultures- death

9. kargi turkey/ bird mountain- death

10. Sabah Mountains / This doesnt fit the pattern because it means morning in modern Turkish and Arabic, but this may have been converted due to Islam as pagan. So I have added it in.

11. saba in bonaire

12. Ulu or Ulur Australia- Great Powerful mountains

13. Aksu Xinjiang China / Holy water mountains

14. Akcay Turkiye / Holy water mountains

15. Kulu Norther India / K was added on after to hide paganism, because Kulu means servant of God, it was most porbably Ulu -Great powerful mountains- human sacrifice

16. Kulu Konya Turkey/ K was added on after to hide paganism, because Kulu means servant of God, it was most porbably Ulu -Great powerful mountains

17. Adamawa Nigeria / Religioius / Man and eve mountains or adam and eve mountains with their pron

18. Bou zizi (phonetically identical to bu sizi) Algeria / This pounding pain mountains

19. Ulu.dag Turkey / Great powerful mountains

20. Uri-switzerland - Dark Fairy Mountains / Along with Alp -It is the newest phonetically sounding which is very interesting because you can hear the overlay and has indo european in it.

21. Alp Switzerland / higher power aka Destructive mountain- It is the newest phonetically

22. Taban Hungary / Inhabitied areas older than Neolithic times, protected valley means tasty palet for feeding.

23. Gdanask Poland / Town of Nest has encountered overlay. But lets look at regional words it is close to El.bag (their orchards) and Ol.stzyn (Ol= death mountains) and south of kashubia - fits all criterias descriptive.

    Kas is found in several places

    I.   kas Turkey

    II.   Kashmir are abnormally excessively steep hills today kash means eyebrow.

24. Ala (phonetically pronoucned the same as it written version silent g- Agla) daglar Turkey / cry alot moutains.

25. Karadag Turkey / our darkest mountains

26. Chugauch Alaska / - moutnains are very hungry cok ac.

27. Aonach Scotland / 10 a stardard shaman prayer format, means the ten is hungry

28. Aonach berg scotland / " **very important explain in detail.**

29. Masada Israel/ **Means** at the table feeding them to Spirits, **the dead**

30. Nasarawa Nigeria / Evil eye mountain nasar awa

31. Buyuk Agri Turkey / Big pain mountain

32. Kucuk Agri Turkey / Little pain moutains

33. Kuskokwim Alaska / My bird sky mountain- death

34. Kechikan Alaska / **The blood of diety** feeding the gods, **sacrificed goat mountain**

35. Belam Central America / **Most** lethal **mountain**

36. Sahara, North Africa / look for land desert

37. Nazar.eth |Israel / Mount of evil eye *** see series 2 Religion of Ute / where the concept of eye came from, and how it transferred to the religions we have today**

38. Kalahari South Africa / Lasting pain moutain

39. Sonoran Mexico / **Last place of visit defined as your place of** death mountain

40. Atacama Chili / The eye that stares moutnain

41. Kyzyl Kum Central Asia / **Rust deserts**

42. Gobi Central Asia / Center of Soul mountain

43. Taklamakan Central Asia / Flipping to afterlife

44. madagascar / the island of force today muscle in anatolia - similar to caracus venuezuela prounced the same karakas south america dark force drop m = ada. kas (gas) kar

45. Zaragoz spain / The eye that gives harm, **mountains or peak referencing lighting**

46. Karakum Central Asia / Dark mountain **converted to dark sand desert over time - indicator of time put icon**

47. Dashet converted from Deschet Iran / Killer lethal mountain converted to sands 30,000 year process

48. Caracas South America (kara.kas) / The force of dark

49. Yukon Canada / The weight of ten mountain shaman prayer format yuk.on

50. Utah (ute)/ The bird tweeting moutnains- feeding them

51. Yucatan Mexico / Moutanins that push force

52. adaklu ghana / adak human sacrifice like alaska other region kumasi ck

NOTES turkic speaking tribes battled between themselves for centuries when no borders existed, we dont understand that today.

they were hated part of our history- because they remained very unchanged when other regions were developing into settlements

lollards England, gypsies romaina, puervcaata spain (pig leaders), natives, i believe these categories of people / groups are linked pertaining to part of the 500 pockets of indigenous spread out globally

Rebuttals

There is no question there will be rebuttals, this is fair. I mean humans constantly associate language with a certain region and the mind gets confused with any type of educational enlightenment that breaks a perception. Having utopian beliefs means the elimination of borders and being sustainable Globally

Facts

we have a 1 million year gap between controlled fires and more advanced settlements.

We have during the lower Palaeolithic periods in Korea evidence of our earliest humans 1/2 million years ago.

We have mammoths a staple food crossing the Bering during around these periods. Understanding there was only 500 pockets of approximately 100 people totalling 55,000 human proto types confuses people in Land mass 1.

This is exactly Birtur's era, I call it the missing link. When the transfers of dialogue from the animal kingdom incrementally built and our earliest humans developed even more complex language.

Lets break perception Aztec and Inca became Spanish associated only because of Conquest. CARAQUE investigated was originally carakawa. Que is latin a suffix,

Hypothesis Journey 1

Why is this 55,000 number so critical because the source of crossover regarding people were predominantly the same blood line that has confused scientist with the number of people who have actually cross. check

Hypothesis Journey 2 more recently Aztec / Inca's

1400 Chinese explorers might have dropped off people they dont like explain / pyramids match those to Korea and china

article indicates only a handful DNA wise match to all the natives in the americas because majority of indigenous got killed with European explorers 1 probability from 20 million on north and south to 2 million remaining or same close family line 2 probability age.

regardless if we ignore all the politics behind history language gives us a different account.

Acccording to prof...xxxx .Written proto turkic (Tamgas) started in Korea which would explain Birtur descriptives in the central Americas and south americas where pyramids are located.

My objective is to show patterns only.

Some mountain names might run into error category mentioned earlier but there are three strong arguments why in general Birtur classification is correct for the descriptive names in the Americas fix the rest

1. Firstly we are working backwards from land mass 2, where there is a dominance and clear pattern of Birtur classification. Whether that being native tribal names (Anasazi), shaman spirit / diety related names (kechikan) or regional descriptives (Tulum)-there is a pattern.

2. Yakut is adaptation of thousands of years to extreme sub zero cold temperatures and to cross the Bering Strait you would need not only knowledge of the topography, plus the passing of historically known hostile tribes such as Mongolia

3. Others way of crossing over to the America's is also possible but my counter arguments will be in book 2 under the global history section and why that probability is slim . Let's remember we are referencing thousands of years prior when only chiselled rocks existed.

How do I know it was that long ago?

   ➢ As per article..xxx ...all natives are linked only to a handful of people from Landmass 1;

- ➤ Birtur is based on a comprehensive comparative with modern Anatolian Turkish or Turkic;

- ➤ I had to use phonetics to decode; and

- ➤ The words are coming out with basic different vocals, not refined and developed like we have in modern languages today. please see webpage www.xxxx.com

Interesting Case Studies

This is 7 fairy mountain in China

Let's do a reverse and connect the dots. If we are to translate this to our Birtur Descriptive format - it means 7 Peri-shan. What connotation can a word like 7 Peri-shan mean.

> Today the word seven in Turkic has two meanings the "actual number" and to eat

> Peri-shan on the other hand has two meanings fairy mountain or to make perishan (perisan our s =sh)   Perishan is an entire descriptive mini sentence. A force that left us feeling destroyed.

Can the etymology of these words be derived from Central Asia?

> If we look at Switzerland the cantons also means fairy but this time in Farsi our newest mountain label means URI / phonetically the word Alp is new as well. This is from an ESL perspective of using different vocals to pronounce the word Alp. Deducing Europe is the newest mountain labelled region. URI was done after Farsi overlay and using more advanced Turkic the word Alp

Can you start seeing a pattern of where 7 might have originated from in Indigenous Turkic and that these mountains were destructive forces for our earliest ancestors.

The Case of Ten

Ten or Birtur translated is the word ON, ON is a shaman prayer format of raising all ten fingers to the sun. Palms getting heat, it is also descriptive.

Today the word 10 may have come from Tengri which means god, all ten fingers for god. Visualize our earliest ancestors not understanding their fingers and looking at them for hours.

Number 10 is found in Chinese children folk tales - the ten suns.

We can also find on.ach Birtur translation - the ten is hungry in Scotland.

The Word Wool Today

Vuul is indo European and means mountain deities of sheep gods in Hindi but phonetically vuul is equivalent to wool

we then find Yunnan mountains in china / Yun translated is wool in Turkic, there are several Yun points out there all mountains. Ketchikan is Birtur descriptive sheep blood mountains in Alaska.

Limitations In Errors

Can I be wrong absolutely but these patterns are consistent with various points around the globe and our objective today is to show the consistency in patterns across the landmasses.

I believe this would require years of research to complete with several collaborations between universities across different countries.

Publication 2 will show difficulties in research under the chapter called distinctions.

Birtur is descriptive photography to the past, the study of these descriptive words and the origin of some words give us historical information. What are earliest ancestors were like. I call this the missing link thinking outside the box to get information on our prehistory.

Once upon a time, for a very extensive period, there roamed on earth a few thousand shamans they left behind a gift for us. It's called X marks their spot, several regional name globally of their dispersed pockets of settlements.

LANGUAGE THE BUILDING PROCESS FROM OVER A MILLION YEARS OF SHAMAN HISTORY - DONT BELIEVE ME NO PROBLEM ITS IN THE BIBLE.

From Animal to Hominid

Stage 3- The Transfer to Language

The question then becomes when did this convert over, on our human timeline, to basic dialogue. I mean if we look at the family of Niger-Congo Languages in the South African regions, those clicks have now been incorporated into full language. I believe the answer is more unusual then some can imagine.

I simply researched when humans or its predecessors greatest technical advances came in, which would have been the discovery of controlled fire. According to J.M Roberts book on prehistory and first civilizations he illustrates " that many scholars based on newer evidence in the Transvaal area believe that hominids were using fire well before homo sapiens even came into the picture.". If we were to debate this, then "many scholars", is not a uniform agreement to suggest language occurred over a million years ago. But I am on the side of the scholars who do believe fire and language came approximately the same time, and that it existed way before the existence of homo sapiens and here is why. Like all living things that were born and evolved incrementally from micro organisms to large mammals, so did the languages of earth. Significantly language was a developmental process in our history. Even variations in the first grunts of early humans had to have been understandable amongst the groups that communicated between them. It is just unfortunate that we as humans don't understand the various tonalities of vocals that each noise, screech or grunt signifies. Lets have look at an example of our timeline.

It illustrates where language transferred over from communicable grunts to full dialogue in humans. It gives us the basis to where language started developing.

Time Chart (4 million years to 100,000 BC).

4,000,000 BC Appearance Australopithecus	3,000,000 BC	2,000,000 BC Homo Habilis (Tools)	1,000,000 BC Homo Erectus (Fire)	600,000 BC Neanderthal	100,000 BC - onwards Homo Sapiens complex language

The colour coding for the boxes Homo Habilis and Homo Erectus are shaded in the same because I would like to add another interesting fact as we try to pinpoint and evidence the undocumented portion of our language origins. It is also the reason the arrow is not fully on the Homo erectus family.

Tools were found in Ethiopia which are the oldest to date at about two and half million years. They are called pebble choppers all selectively prepared. As basic as these tools seem someone/ groups of people from our past took the time to carefully chip away at these stones in a very thought out manner. Very different from certain animals in the animal kingdom who randomly picks up tools to aid with hunting. Some will argue that such exciting evidences by academics as fire and tools are not sufficient when it comes to evidencing language. I beg to differ.

The introduction of Birtur and our Pictionary descriptive index in the back along with nomadic patterns, graphs and diagrams in later chapters will substantiate this further. In addition, what is better evidence than the Harvard University DNA Migration Map as our guide anyways?

Then there are the no outside contact, indigenous tribes of America's, which are a great look into the past where all their products are not only biodegradable, with constantly changing makeshift settlements but full dialogue amongst the members themselves does exist.. To the point that these tribes, their dialogues, cultures, beliefs is a peek into what life was like 100,000 years ago and should be compared to our origins out of Africa.

Introduction of Birtur - Early Humans

Stage 4 - Start of our Pictionary Linguistics

Close to a million years ago humans discovered fire, the hypothesis to follow is that it is around this period that the start of conjoining, single syllable words, started forming.

These single syllable words were facial expressions, to surrounding events, converted to sounds. These are called reactionary responses. Therefore what are visually associated phonetic noises? They are the commencement of the standardization of sounds as a component of starting of speech. And lastly to its slightly more advanced version which are full detailed descriptive words, known as "Pictionary visuals" of the first words of earth.

For the purpose of this publication I will call this particular Pictionary linguistic, Birtur. Its name is translated in modern Turkish as "*first type*" and these newly discovered patterns should be dated as an earlier version of proto-Turkic belonging to the agglutinative family of linguistics. Also described as the language of the first clans that grew and broke off, from their original pods of a few hundred hominids somewhere within the African continent. It is our attempt now to classify and organize these first word visuals which will also help in tracing the migration of our earliest nomads.

In the index page at the back of this publication you will get a list. It is our attempt to categories these words to illustrate patterns associated with early human dialogues. Here is a list of the first noises derived from those descriptive words.

format rechange

An (*Un*) like Ana or Anu / Mother	Si - pronounced like a hiss from a snake/ water
Ag (*extended a*) like Aga / Head of, normally tribal leader	Uu pronounced
At - (*Ath*) Ata / Leader	Ac (*ac*) / Hungry
Bo (beuh) Bocek / Bug	Yan (*yan*) / Burn
Ka (*ka*) kara / Dark	Yi (*extended eee*) like yilan / snake
Ku (*koo*) kus / Bird	Can (say *john quickly*) / Life
Sa (*sah*) sara / yellow	Han (*hun*)/ inside a variety of words ie. shan mountains
Su (*Suu*) / Water	Tan (*tun*) / higher being

Linguistically why classify Birtur a completely different category to that of proto-Turkic?

For the following reasons:

16. These are Pictionary words starter words, what our environments contributed in terms of development of language in early humans;

17. There is a direct correlation between these descriptive words to immediate surroundings and regional names, or actions;

18. These are single syllable or very basic conjoining words;

19. Some are decoded words phonetically, but not necessarily found in its same format in modern Turkish or Turkic;

20. They may have transferred over linguistically but because of its age they may have different meanings to that of the Turkic family of languages today;

21. They are consistent patterns or grouping of words that are associated with absolutely *"no archaeological find, or written portion"* of history but rather the study of language;

22. And lastly for educational purposes. To illustrate its phonetics, word order and linguistic contributions to a variety of languages we have today.

Birtur Classification Criteria's

What will be our methodology? We know that Birtur is what our first humans saw topographically and its association to the development of language. We also know that these regional markings are predominantly found along coastal lines, mountains / desert ranges and some islands. Evidencing a complex linguistic of our earliest nomads and their preferred migration patterns. But there are other words from the regions of the first civilizations that should be incorporated as Birtur, why? Because they contain 2 or 3 smaller single syllable that have been conjoined and are actually an action versus the animate object it defines today. These words also fit in with our early ancestors.

A Simplified Summary:

FACTORS	CHECKLIST
Does the word correspond to its period?	Approximately 1.5 million years old to 15,000 BC- where the stagnant period starts coming to an end for singular linguistic, due to geographical distribution and isolations. Like that of the America's.
Is there substantial evidence?	Are there other closer regional words that match? Do these words fit together like a link? Is it close to an area associated with the first civilizations? Understanding that these were small communities of early humans.
Does it fall into recognizable word classification?	Do the words match possible behaviours to that of early humans? See the case study on the word Dokkum.
Are the locations consistent?	Are the location consistent to other areas where these words exist, predominantly along coastal lines, mountains and islands (near land)?
Do we recognize when the language transfer over?	Can you see slight modifications in words done by the local languages, this is in terms of phonetic decoding? Meaning direct modifications a language of origin may have done?

What's its definition?	Does it fit the criteria as a descriptive region, or a descriptive action?
	Now Let's Play Pictionary

** Please note similar sounding words of Turkic origin may exist but this may not necessary fit into Birtur's general classification criteria and could be completely coincidental. What we are looking for are consistent patterns across the board of early human dialogue.

ACTION Descriptive

1.   The case study of the word KAR-BUZ (Snow-Ice)

So I walk into a Afghani store and there is this odd shaped orange fruit, with the label Karbuz from Pakistan, I start investigating because Karbuz decoded phonetically means snow ice. Why on earth would Pakistan, with a completely different dialogue have a word that means snow ice. My investigator hat comes on and I start researching. This is when the world had no refrigeration, a striking similarity comes to mind. A flash back in time to a remote village in Turkey that I visited where locals still bury vegetables as a natural refrigeration process, under piles of snow. The beauty about first civilizations or the Indigenous people in the Amazon is that very little has changed, other than time.

We have to be aware that early humans did not have ownership of anything, even to something very basic like a label for a fruit, meaning they did not care that an apple was an apple, they cared what they did with it. In this case carry it to the mountains in weaved baskets and bury it under snow and ice.

Karbuz today is linguistically transferred over to the word karpuz (watermelon) with zero affiliation to what our indigenous forefathers did with the actual fruit. Karbuz therefore is a word I classify as Pictionary descriptive in this case it is descriptive of "an act of doing something" of our early ancestors.

REGIONAL Descriptive

The case study of the word Dokkum, in the Netherlands

2.   Here is where it gets a little complicated, my job as the researcher is to a) show what we have hidden historically, b) to evolve. As I spend hours looking at an atlas I am attempting to categorize these descriptive patterns and this is why I have used this particular region as a good example. It also illustrates some of the difficulty I face. I am not working with technology for carbon dated artefacts I am swimming through mounds of word and gluing patterns together while I investigate.

If we read this word in modern Turkish today Dokkum in the Netherlands, sounds like the word dökum, which means to cast metal. The problem is the act of casting metal does not fit into Birtur criteria's for age, since I believe these clans were way too primitive.

So I go to round two of investigating which is to listen to someone Dutch pronounce their regional word Dokkum. Out it comes the silent T in their pronunciation which is equivalent to our word Tokum. Like a piece missing out of a jigsaw puzzle it fits perfectly into our general patterns which will be discussed later. Tokum means *the act of being full,* now why would the Netherlands, near a water tributary possibly have a word that means full this is where we go to our section on Nomadic Patterns in Chapter xxxx for your answers.

As a little introductory clue however the concept of being full was early humans way of disposing of decease, via birds and animals. Understanding that this was a regional point where feeding human remains to birds or animals was common in periods when shovels did not exist for early humans.

Using the Criteria's above lets dissect the word Dokkum.

e) **Does the word correspond to its period?** Birtur dated era is anywhere between 1.5 million years old to 15,000 BC (prior to overlay something that be introduced in the next chapter), feeding of animals yes it fits next √

f) **Is there substantial evidence**? Netherlands are close to Finland which has a confirmed 2% of Turkic in its language. Not Anatolian Turkish, but rather Indigenous people's Turkic. This is the same analogy as when Judaism was found in Ethiopia, in reference to the variation of people with a common denominator. yes it fits next √

g) **Does it fall into recognizable word classification?** By breaking it up into the single syllable in *Dok*.<u>kum</u> or Kum or Dök this belongs to the agglutinative Turkic family of languages, found throughout central Asia and Anatolia. Very common words, yes it fits next. √

h) **Are the locations consistent?** In this case yes this is along a coastal line. √

i) **Do we recognize when the language transfer over?** Based on the country being Europe very possible Neanderthals period. √

j) **What's its definition?** Does it fits criteria as a descriptive region, yes because it means the act of being full and we have similar words in the America's which illustrate a different word but same meanings. See Nomadic Patterns. √

Stage 5

Does anyone know what a ululation is? That is the high pitch tongue thrill noise that is used in most celebrations across Africa, Anatolia and the Middle East today. What is the significance of this odd type of noise that is so common for the call of celebration in these regions, well it is better define as the method our ancestors used to initially call one another, or alert that something was going on, done over the thousands of years of interactions.

Ululations is what I call our stage 5 component of speech and ironically it is associated with most historically Ottoman controlled regions aka regions known as first civilizations. Why am I saying stage 5 component of speech and not first or second, because we can deduce the following:

a. Regions in Africa already contain patterns of similar vocabulary words (will be discussed later) that are older;

b. The complexity of the groupings of the words la-la-la-la in a rapid sequential format (uses different vocals for speech), this is my ESL training now, is another indicative that the existence of language already had been established;

c. Groups of clan may have started distinctions and created that as a form of calling, to distinguish themselves during attacks.

d. The newly discovered Gobekli Tepe (Sanli Urfa) was a known regions to have temporary makeshift settlements. To create this structure, very organize complex communication had to have existed.

Now let visualize, and imagine we are on a mountain by a megalithic structure like Gobekli tepe in southern Turkey, our earliest ancestors would ululate to call for a variety reasons. Today that call is still used but predominantly for celebrations.

The critical importance of the Agglutinative Language family

Why are the family grouping of agglutinative languages so critical in our attempt to chronologically order languages.

Agglutinative languages were the first dominant grouping of languages of earth, and the most naturally spreading. "Most word are formed by joining morphemes together" or easier suffix additions by gluing single syllables. In general the family of agglutinative languages are easier to learn, especially those associated with Birtur, because they were created with sounds to phonetics, incrementally done over time. Remember the concept of knitting in the acronyms page. Agglutinative languages are the yarn to the various outfits that have been produced over time.

These words are naturally developed in our linguistic history and require less memorization. The critical importance of the Turkic family of languages is that everything, absolutely everything said can be written out phonetically regardless of dialects. Remember nursery rhymes? The start of language was the start of basic sounds to the changes of single words, proto Turkic for example has: at, bat, yat, kat, hat. These types of words transferred over to other family groups of linguistics with different sound but same order as follows. In English:

at bat cat that and hat

Remember Marco Polo and his travels across Asia and the Indian subcontinent.

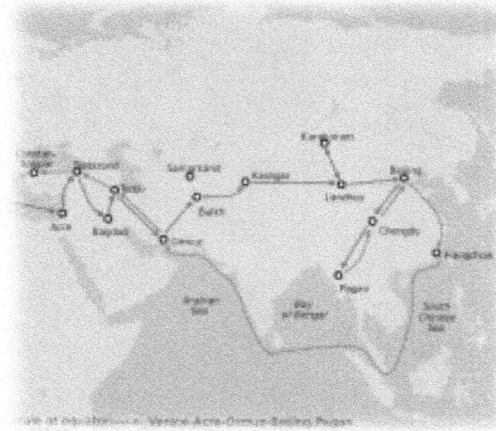

In the 1300's with his basic Turkic he was able to travel and communicate with many of the Asiatic tribes that had varying degrees of Turkic dialect without too much difficulty. No different than how our nomadic clans travelled 1 million years ago as they started developing their own regional dialects, which occurred historically. The simplicity, logical order and the ease of noises to word association evidence our theory further.

The question I pose is what occurred historically that brought on such radical changes in language? Now let's debate?

The next paragraph is called the "first language" pondering, these dual perspectives through the publication is to aide us on being more tolerant on the topics presented.

People of Faith - If we feel our hearts pound we have a purpose, religious books are there to guide us. In this case if the Bible indicated, "that humans had only one language historically", this account should be investigated further.

Atheist - They are after all stories, maybe we should take these historical stories seriously, because they may actually incorporate events of our past.

Visual Summary

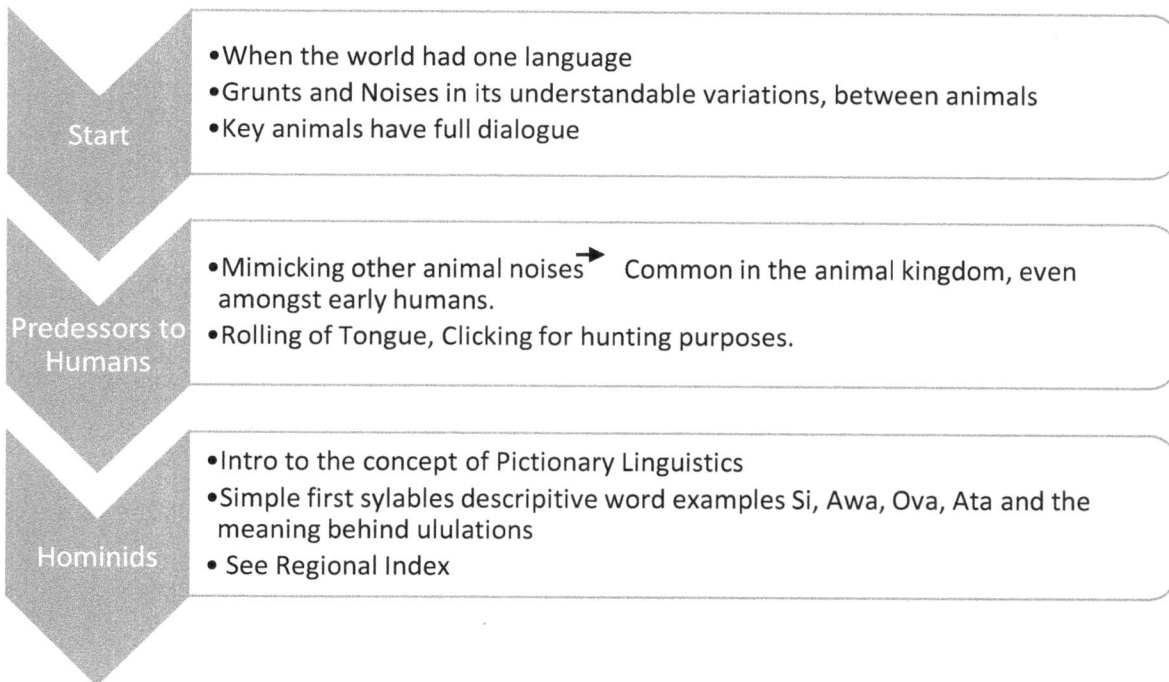

Start
- When the world had one language
- Grunts and Noises in its understandable variations, between animals
- Key animals have full dialogue

Predessors to Humans
- Mimicking other animal noises → Common in the animal kingdom, even amongst early humans.
- Rolling of Tongue, Clicking for hunting purposes.

Hominids
- Intro to the concept of Pictionary Linguistics
- Simple first sylables descripitive word examples Si, Awa, Ova, Ata and the meaning behind ululations
- See Regional Index

Our common words of this globe – separating dialects from linguistics. Was this genuinely not knowing or 2000 years of social control of our first indigenous people globally.

The Connection of our words pull out form these countries	Birtur its Proto-Turkic format	Pictionary Linguistics & Descriptions
Alaska Kuskokwim Mts	Kogruk, Kayuk Oogruk (seal) Alas / <u>Kan</u>	A name on bot hsides of the Bering Strait Canoe Alas - Kan (blood)
Armenia	List of mountains in Armenia	Are proto-turkic put list here check old maps
Ethiopia	He-bes / stan	Place of Trade- major trading center. scales with 5 seed that act as weights measurement to give 1/2 pound explain- coffee trade explain *important
Bulgaria	kulata	Leader of soul -god servant

Sudan	Su-dan	Land of Water *
Malta	Drop m you get a classic	alta
Uganda	Remove u	Kan'da at violence place g and k are interchangeable
Dondo	Aka / kasowa ck in map, dondo	Frozen, ova of birds, aka
	Near shanghai utuan	To tweet
ecuador	Altar, ata-cazo, ana	Slight modification to spanish
Syria	Su-riya same as bia in arabic	In modern turkish riya means two sided may have meant something different then *
Russia Don River	Don	Frozen river
Canada	Don common	Don valley Don mills etc.
Egypt	Misir	Land of Corn

Kom-el Kanater	Kum El kanatir	Sand will make you bleed Base words of a lot of hieroglyphs are proto-Turkic
Afo		
Italy	PROF WORK	
Spain		
Yemen	Yemen	Religion- defined as
Palestine	Fil-Astin	Hanging of Elephant
Israel-Today Jerusalem	Masa-da Mountains Yerusha	Table - genuinely looks like a table I am the people of Yerusha add the liyim, Yerushaliyim this is also proto-turkic meaning "I am from the clan Yerusha"
Southern-Turkey Iraq (Kurdistan)	(i) Kurt Alan, and the other side of Firat river (ii) Kozluk	Region of Wolves / Region of Birds

India	Hindi-stan	People of Hindi - bird people
Calcutta maybe new not new but newer	Historic name Kalik-ata more due to historical conquest but the word ata is standard	Region of Kalik leader - newer due to integration
Turkey Central Asia Turkic	Clans of Hindi Kus	Indigenous bird people- the region of landing birds
Kamchatka	Cam-yat-ka	Sleeping place by Pine Trees
Kandahar	Kan-da har = Kan dagi	Blood- similar phonetically to blood mountain
Mongolia	Mongolia was named by the British Every single desert or mountain range in the region Altai, Taklamakan, Kunlun, karakum, kyzyl kum	Land of flowers Magnolia Is proto-Turkic Rust colored sand
Korea		
Siberia	Su-beri and Deniz-ova	By the Water, Inland of water
Siberia	Si -leri	By the water

Kosovo	Kosovo	My guest this because Russia / Serbia changed the ovo from ova it is koz-ova Inland of Birds area
Guiana	Gui-ana	Certain words have not changed historically and more research maybe needed but gui-ana today guyana is another of them the word ana may not be a coincidence. research nazca plate why that name
Mexico Pyramids Yucatan Peninsula Sonoran desert	Find them Yuc-atan Son- orana Belam Tulum Kuku /can	 The one who powers Last stop The one who gives destruction The white outfit of death We make noises of a bird in this format kuku -can the noises of bird and their soul
Canada	Kanada / Kanat-ta	Protective canopy of wing

Items with astericks (*)	Reflect regions of the world which were once abundant with water and / or agriculture	Human Impacts of deforestation

Countries	Birtur its Proto-Turkic format	Pictionary Linguistics & Descriptions Dictionary format
Alaska	Immediately west of the Bering Strait were all Asiatic tribes that spoke Turkic	That land bridge existed approximately 25,000 to 75,000 years ago- memory ck again
Alaska	Alaskan mountain ranges	Alas - Kan It is my assumption that the Kan word is indicative that primitive humans started acknowledging death and higher power. Most Kan words are associated with the top of mountain, where bodies where fed to the birds as a form of disposal.
Alaska	Kuskokwim Mts (Kus / Kok or Gok / wim)	Kus is bird, Kok is root or Gok is sky the wim is possession. My Bird Sky

Alaska	Ulukhaktok Uluk / hak/ tok	Ulu - a high place
Alaska	Kobuk	Means shell region found near water
Alaska	Kululak Bay Kulu / lak	Kulu servant to God
Alaska	Yakutat Region of origin, which is the west of the Bering Strait	Yakut Today Ruby
Alaska	Naknek	Very similar to Inek animal and Nanak
Alaska (Mountain)	Chugach Mountains	Turkic is so logical in format that what sounds come out is exactly how it is written. Keep in mind foreigners probably spelt out most of

		these words. When I read this phonetically, this sounds like cok ac (pronounced a-ch) in modern Turkish which means "very hungry mountains". Concept that these mountains probably fed a number of deceased to birds.
Alaska (Mountain)	Tanana	Tan-Ana Mother God as Tan is associated with Tengri, Tanri. Ana is standard in central asia.
Alaska	Koyu<u>kuk</u>	Koyu-Kuk - Kuk being Gok (dialect variance) Black Sky
Alaska Yukon Hills	In the 1917 map there is an accent on the u which is indicative of emphasis.	Yuk means the weight of the God- and its topography being the hills. First burial

		formats of our primitive ancestors.
		That same format in Mexico comes as Yuc-attan the one who throws weight again
		Mountain ranges. ON mean the weight of sun on 10 fingers
Alaska ck	Kaiyuh	Pronounced Kay-yuk Mountains of rocks modern Turkic has Kaya
Alaska	Karluk	Kar-luk snowy region
Alaska	Many others in their agglutinative formats	Papka, Ekilik, Kaguyak, Kashalak, Kokotik
Armenia - mountains	Kapan - use to house a small jewish community.	The one who takes away life
Northern Chile	Ata-cama desert	Which was probably once a mountain-we can deduce it

	Leader of eye	took approximately 25,000-50,000 years to turn a single mountain to sand based on when these clans moved across the Bering strait. research
Congo	Kayna	To burn
Congo	Kutu	box
Liberia	Zorzor mountains	Hard hard emotionally mountains
	Kakata	leader
Liberia	Bird artefacts	
Ivory coast	Bird artefacts	
Afghanistan	Kabul, kan-daha agir, konar	Accept, blood mountains weight, landing of birds
Tibet	Kunlun	Sandy area
Siberia	Su beri	By the water
Siberia	Denizova	Inlet of sea

Hindikush	Hindi-kus	Bird clan
Morocco (Mountain)	Bu zizi	This pain bu sizi
India (Mountain)	Kalanag	Lasting karma
Himalayan (Mountain)	Hima-al-aya	Aya is moon
Turkey Bogazkoy	Yazilikaya	Aka Written rocks investigate age of name is it new
Alaska	Koyukon	Koyu is dark kon could be blood kan, but kon today is to land.
ISRAEL	Yerushaliyim- old name of Jerusalem.	I am from yerusha clan. The ending liyim is proto turkic for ownership of a territorial clan. Use commonly even today a good example. Istanbul-liyim.
	Hosanna	Beautiful mother
Montana	Mother Spirit / mountain	Mont Converted from spainish Mont-ana
USA native	Ute	Bird tweeting clan

USA native	APA-CHE	LEADER SPIRIT
USA native	Karankawa	Blackhead or black sky
USA native	Tahana	Mother god tanana
USA NATIVE	Yuma	What wool gives 3rd place goat divinity
USA NATIVE	Kalapuya	The fitting part
USA NATIVE	Acaama	Hungry mother- so much suffering
USA NATIVE	ANASAZI	MOTHER OF PAIN
Austria	Kara-wan-ken	Birtur mix farsi
	Asul-kan range research	
	Kokannee	Gok sky
Bellize	maya	
	Talamanca central salamanca	Tribes seperate linguistics ck
Morocco	agadir	It was heavy
Morocco	Tan-tan	God-God
Morocco	Fark-ana	Different mother
Morocco	aklim	The mind

Morocco	Zir-ara	
Netherlands	Dokkum Makkum Berilkum kollum	Casting of could be iron or steel We cant see outside the box because of our genetic trait to apply distinction or color code humans. That there existed one pod that produced 20-30 clans in a stable format for thousands of years communicating. when pop explosion occurred the need to self preserve and create distinctions got enhanced.
Andes check did late doesnt seem right	Ata-cama Chincha-suyu Colla-suyu Hatun Canar Sara-guro	Ata is quite common and means Leader ....type of water ....type of water female from soul Yellow....

Finnland	There linguistic also belongs to the agglutinative family-reindeers etc how indigenous are they	Research Tu'run, Keski, Kok-kola (sky or root), Kaskinen, Kalajoki, Kuotane,Oulu, suomenselk
Brazil	Maraba	hello
Peru	Mont ana	Ana classic word
Finnland	Turku abo	
Mexico - find article backup	Tehuantepe/c Gulf Indicative of Region Toluca	Tehu -an-tepe Expression of sadness Tehu-tepe mount Uca to fly
Mexico	Cancun (same region)	In spanish was Kan-kun a birtur word the sand made of blood probably referencing clan fighting
Mexico	Yucatan coastline region Majority in Mexico is in the Yucatan peninsula if there was many indigenous people	Yuc-atan The one who throws weight-religion connotation

	governments may not have change the names, no need felt for distinctions. check or was this a large clan in one area only	
Mexico	Pich	A derogatory form of wording to announce a fatherless child, but when we study amazonian natives we learn that fatherless children are seen as evil spirits and are removed from the clans and put to certain areas to get killed. There may be a connection as well to the indigenous tribes in yucatan region.
Mexico	Sinaloa	Sinal-ova
Mexico	Kanasin	Let it bleen Kan -a -sin Let it bleed

Mexico	Bacalar	Bac-a-lar Smoke tops
Yukon	Athabasca	Atha-basca ata is different soul Where c=k in Turkic (baska) Means our Ata is different
South America	General	Ending in ana mother Santa-ana, Quitana indicative of a slight change in wording. Ana is primitive- changes from Spanish
South Africa	Kalahari	Kala-agri staying pain
El salvador	Usulutan	Usul-utan- the one who makes passive most likely reference to higher power Quiet -be ashamed
Mexico	Oajaca	Oaj-aca ...ck
Mexico	Tapana	Tapana the one who worships

Mexico	Maya Mountains	
Mexico	Mayahan	Maya-han
Mexico	Belam	My deadly one
Mexico	Zanatepe/c	Could be Kanat-tepe ck
Mexico	Tepich	Ch is pronounced s in Turkic Tepis- rapid movements
Mexico	Kantunilkin	Kan-tun-ilkin Blood ..first of the family plus its my last name deal with it. Also tell him about gotland hahaha
Mexico	Sula	To water
Mexico	Tulum	White wrap for the dead
Mexico	Sian Kan'an	Sian Kan'an.... Canan To give God
Mexico	Nohbec	Nobet -to guard
Mexico	Ekbalam	Ek-balam (belam) The one who gives evil

Mexico	Many others in their agglutinative formats	Cuernav-aca, Co-ba, Saban-cuy
Mexico	Hunucma	Hun (leader) dont fly away
Mexico	El-tigre, el planchon Many others in their agglutinative formats, el cruce	Arab influence on spain during overlay indicative of integration
Mexico	Coban	The one who follows herds
China	The following are regional markers for China	China was also the first to implement distinctions and name changes approximately 500 years ago. To evidence this all we have to do is look at Korea right next door where proto-Turkic literature advanced. Professor xxxx reference
Korea - find professors work - where turkic linguistics advanced	Chang dok palace Koyongbok	Can dok - life spilled palace Koyun bak explain
China	The Chinese pronounce our Z or S with an X	That most of the Han words have an S in the front to do the distinction for what they

		felt where attacks from the Islamic crusaders at the time.
China	Faichan Kangri Mountains Faiga-Kan-gri	Terrible blood mountains
China	Yun Mountain	Wool Mountain
China	Cangyan	Can-yan our g is silent Burnt soul
China	Das-han and Mount Hou	Sound phonetically close to Tas-han Stone warrior and Mount prayer Dua Let's remember there are some dialect differences and some changes with the creation of distinctions 500 years ago.
China	Muztagh Ata	Mutac- ata could be Dependent Ata

		Ata is one of our global marker core words
China	A few listed mountains as Kangri or Gangri	My guess Tangri / Tengri God or Blood
China	Chokusu	Cok-su Lots of water
China	Yunnan	Today transposed to modern Turkish as the people of Greece
China	Yilan	Snake
China	Xian	Ziyan -City of waste my guess lost of life
China	Shanxi	My guess Sanci the area of pain referencing torture
Greece maybe new	kalamata	Ka -lam-ata newer due to integration
	illeriyia	Phonetically means advanced

Kuala lumpar	Check	I forgot city
China	Kagan /bulak	Bejing's historic name Kaganubulak -means King in Turkic and Bulak means large settlements.
China	Bayan	Female
China	Dalou Shan	Dagli- Shan Mountain Shan ask someone asian
China	Bayanhar Shan	Bayan-har shan ask shan father
China	Gansu	Kansu blood water
China	Many others in their agglutinative formats	Tumen, Duyun, Kali, keshan, Haiyan,
China	Jiangsu	Cihan-su our J is a C
China	Da Yunhe	Mountain of wool Dag (g is silent) Yun wool check
Croatia	Dinara	Search for god

Croatia	Uc-ka	To fly
Croatia	Kozjak	Koz was a bird suffix seems to have been modified
Croatia	Hrvatski	Hrvatkoy research
Estonian	Eesti	To blow wind
Croatia	Zumbera-cka	Storage today piggy bank Word stands out.
Balkan	Bal-kan	Sweet blood honey
Croatia	Papuk	Shoe leather
Croatia	Vidova gora	Vid -ova gora switched to gore ---location of ova check
Kualalampur	Kokurikulum	Gok-uri kulum ck for more I am servent to one who makes noise from the sky
Bejing	Historic name han / Bulak	Side -settlement
Yemen	Sana'a	For you
Yemen	Dhamar	vien

Yemen	Sayyan	The one who counts
Yemen	Sahar	
Yemen	Yarim	half
Yemen	Madiyya	
Spain	Azahar	
Spain	Vilanova	Could be yilan-ova ova is standard
Spain	Zaragoz	Zarar-goz ck if mountains or saragoz
Evidence italian researrch by professor	Italy spain ck family	
	Italy	Turin, asti,
Spain	Calp	Heart Kalp
Spain	Daurada	Da -ur-ada luck island ck
Spain	Altas	Al-tas
Spain	Salamanca	We say that if someone speaks a certain language labelled in this case we would say he speaks salamanca

Spain	Al-canar	Al-
Serbia	tuzla	A name also found in the modern turkey means to salt.
Russia	Russia did name changes approximately 100-150 years ago but we have some clues. When you see the suffix ski it actually means koy in older maps. Koy means village

The same with ovo which is ova area surrounded by water

Example Balakovo ck

Manturovo

Russia had numerous markers of proto-Turkic names. To the point there is a saying in modern Turkey ..."dig the ancestry of a Russian and you will find a Turk" | Balikova and could it be mantarova ck historical name

From kazan to pacific research. all turkic |
| Russia | Makar'ev and Baba' ev | Ev is home |

284

Russia	Ugra	Come visit
Russia	Kuusankoski	Bird san region ck
Russia	Samara	Saddle
Russia	Totma	Dont hold
Russia	Aldan, Baikal, Avar, Kara river, Dersu, ilim, kazan,	Made of red, turkic name, dark river, der-su water, warm, made of iron.
Burma	To twist	
Burma	Check there is another	
Spain	Catalunya	Catal yolunda abbreviated. split in the road
Bulgarian	Are actually known as the Bulgar clans	
Swiss Alps	Alp = Karahman	Defined as warrior for those who climb
Pakistan / India / China	Kara-koram	Black Soot Mountains
United States West Coast	Alta -Great Basin	
United States West Coast	Tul Lake	

United States West Coast oregon	Nehalem	Look at my life in a sad way
	Siskiyou mts	Possible Spitting montains
	arc-ata	
Look at native names coast US		

nebuchadnezzar are several words, ninehev is din ev

water find some examples holy water

thanni / tamil water is than.ni which is after overlay god tan.ni holy water

watergoddess india sudantan forget ck

osaka is os.aka meandering spring water japan

The second is the Birtur word Papuk, it got transposed to the following a mountainous region in Croatia, but refers to hand sown shoes all across Anatolia and Central Asia and can be broken down to the french word pas - meaning feet, by using the first syllable of the Birtur word Pa-puk.

The third is the wan van von Today in Farsi it references region or settlement, in ...german it refe someone established from a settlement in birtur....

period, and the same language transfer period

Conclusion

In the end our human connection to earth runs deep and the only dividing factor that distinguishes us is awareness.

All areas of enlightenment starts with one person merely pointing it out. Now let's move on to evolving.

vatican =bati kan hidden for you only the name is even violent we are evolving???

Understanding History

War is defined as the violent conflict between groups of people. The brutality of war has not only contributed to significant loss of life but has had detrimental damages on human history. The elimination, destruction, pillaging, modifying, of our historical past goes hand in hand with any acts of war that has existed through the thousands of years on this planet. The total number of wars tabulated based on.....quote figure of the number of wars that occurred on earth is over 3900.

Even some primitive humans upon victory would use their new captured slaves to have them grind up the bones of their ancestors to delete the reminders of everything of their past. In saying that history has inaccuracies that hopefully with the information age will come out to prevail as the truth and this is critical for englightment.

✦ Now let's remember history has always been written by the winners of conquest, that until lions have their own histories, the hunt shall always be in favour of the hunter regarding history.

Feedback

➢ Are you someone who may know a story of a region or have caught an error so we can investigate it further or rectify it in a next edition. In the end I am a researcher so part of my criteria's of research was to incorporate a feedback section to limit errors. This section is very important for us from a historical perspective. All information provided may be given to the dying language institute for recording and preservation. The study of historical linguistics provides astonishing descriptions and details to our past. Human objectives should always be to research, preserve and protect all global heritages. This includes language and or descriptive linguistics.

➤ I believe just one descriptive label for a region can be a whole PHD thesis. Please message me at....xxx.history.com, regarding academic information.

something like this

Everything we have learned to start evolving is presented here....

- how color was derived

- the genetics of clans

- the need to support the 2 critical programs genog and dying languages

- clans

- nomadic patterns

- make list

Now we begin the process of understanding the ethics behind why humans did what they did historically and a 100 page proposal for human betterment.

The Venting Zone	Info Email goes straight to the Dying Languages Institute
This is an open forum for anyone who hates Niggers, think Turks are apes, Natives are disgusting wants to run on the street with a national flag because they miss the whole section on, generational social patenting. They are confused and still hate, that they can't make a distinction between academia and politics. They think the authors a local prostitute. they are embarrassed.	Banarasi (Ban.ara.si) is a shaman word for breadless leaf wrap with grains of food & water. this is a perfect example of words that can helps us do anthropological research. Do you know of any

They are not getting the single species concept known as "human". I mean this section is for you.

You can post openly with no name or create your own thread it is called the venting zone it is about channelling inner feelings in a healthy way. This is a Real thread. As long as there is no violence (that is never tolerated) pictures can be posted too...

others? It doesnt matter if it is not shaman Turkic. If it belongs to antiquity we want it. We also accept anything Preserving our global Shaman History., old folk tales, quotes etc. email......have a story

Glossary

10,000 BC year mark- its in the bible Language changes =population increases that point coming out of the 55,000 mark and becoming 1 million people	The period when Shamanism as a culture has started to end and overlay starts. The start of social patenting. Bird religions, settlements, more severe territorial

The concept of MINE DOMINATION	controls, leading to domination type territorial controls etc. Not to be confused with shamanic doctors in central asia or North american native medical doctors. Shamanism references early human living formats, hunter gatherer style and migratory movements.
AD	After Death
Agglutinative languages	The first dominant grouping of languages of earth, and the most naturally spreading. "Most word are formed by joining endings of morphemes together". Agglutinative languages are generally easier to learn because they require less memorization due to its syntax nature.
Animal Kingdom Comparative	
Anatolia	Modern Republic of Turkey and regions
Ashirets	Anatolian word for Clans
Balal	What I have "re-labelled" now as the "transitioning phase" of the introduction of various global linguistics versus what is historically known as the languages previous to Sumerian linguistics.
BC	Before Christ
Pictionary	Birtur is defined as what our earliest predecessors mimicked in the animal

	kingdom, what they visually saw or encountered topographically, projected into first sounds. Its associated era is plus or minus 2 million years prior to 15,000B.C.
Pictionary Classification of Linguistics	Newly *discovered consistent patterns*, or otherwise known as a form of descriptive labelling known also known as "Pictionary linguistic", a proto- type Turkic language, belonging to the agglutinative family. Predominantly found along coastal lines and mountains ranges. Evidencing a complex linguistic of our earliest nomads.
Pictionary Developmental Period	What our earliest predecessors saw topographically and its association to the development of language or the transfer of information phase, in building first words. It is also known as the period of natural development of the first languages of earth in its variant stages, which contributed to majority of the languages of earth today.
Bering Land Bridge Crossing	The bridge used to cross from Asia to the Americas. Collapsed 8000 years ago
Brain Start	
Clan territory	Region associated with aproximtely 100 people.....

Clans or Tribes	A group of primitive people associated with the same family via interbreeding and controlling territorial regions. Can still be found today in regions of antiquity that have not really altered living formats like Anatolia and Hindus Valley.
Clan Sprawl	
English Language	A dominant Indo-European language spread globally, not only due to colonization but with the advancement of technology today. A language most probably born in Anatolia based on the discovery of newly found cuneiform writing in Bogaz Koy Turkey. See Time Travel.
environmentalism umbrella	
First civilizations of the world	
Global Illumination	The concept that through enlightenment and education humans can collectively achieve healthier settings and living environments.

Gobekli tepe	
Halay	
Human Integration (*Direct*)	When one or more cultures immediately following a language transformation due to overlay, integrate.
	It is the cultural outcomes that occur: example
	Tatar Russians (Turki⇨ Russian), or an example of an opposite format:
	Seljuks Turks (Predominantly Farsi ⇨ Ottoman Turkic)
Human Integration (*Independent*)	When one or more cultures not associated with overlay integrate. The movement of people due to slavery for example, the Chinese community brought to Jamaica and their integration.
	Known today as the "*Caribbean Chinese Community*".
Indo European Languages	Born africa
initial mutation	
Intermix Phase	Periods during the start of settlements that created nomadic movements, drastic

	changes and integration of clans due to major environmental impacts. find scientific sources • China Buyo that forced tribes back towards anatolia 13,000 BC , • Black lake (Deluge hypothesis) becoming the Black sea 6,000BC.
Incremental Development	A metaphor word used throughout the publication as incremental steps, with end results being seen only after completion.
Informational era	
Land Mass 1	Another definition for Europe, Africa and Central Asia combined
Land Mass 2	Another definition for North and South America combined
Linguistic Overlay	The concept that with geographical isolation of clans or with intentional human alterations to a language came about, a polyphony of global languages and dialects we have today.
Linguistic Overlay- Human Altered	Is a direct result of primitive humans implementing distinctions in language, in most cases intentionally done via their own language scholars. The emergence of newer more complex and multifaceted dialogues where the language foundations and words are predominantly memorization based, a good example of this is the more complex

	language families, Semitic, Indo European, Sino Tibetan. *Intentional changes by humans in language have nothing to do with various "literary works" out there.*
Linguistic Overlay - Natural	Comes in one format only which is geographical isolation of groups of people during nomadic travels. Meaning new landscapes, generated new words. Example Dodo bird only found in Madagascar- which would have generated a new word. Over time this isolation started making primitive humans gradually change naturally their dialects and lexicons to their new topographies and create other Pictionary visuals of what they historically saw. Landscape changes therefore created new words and added on or naturally changed existing communication over time.
Linguistic History	The study of language history, today also defined as the categorization of language based on their complexities and chronology of development.
Man Who Walk	

Missing Million Gap	The transitional period from the animal kingdom to the birth of early humans, when language existed but we had no records of it.
Mona Lisa Theory	The concept that only one of each species was born to earth. That two identical species at any point could not have been made. A good parallel would be like the impossibility of replicating an identical Mona Lisa painting twice.
Mr. Sunshine T.shirt	Charity program for landscape betterment www.sun.com
Mutation	The idea that all species were born from a mutation
Natural inherent practice	The practice of intermating
One Family Theory	The theory that everything on earth is interconnected via self-replication/ mutations and that only one single female was able to create a mutated species called human, therefore deducing that we as humans are all related.
Our own stans	
Palaeolithic Language	A stagnant period of time of the first language of earth dating approximately 1,000,000 BC to the start of overlay globally 15,000 BC. Homo Neanderthal inclusive evidencing complex language; what was missed from

	our history. See Time Travel for nomadic patterns during this period. Another name for Birtur Linguistics.
Photo-Trek from Africa to the Americas	Photographic snap shot of our first humans of earth globally put together in a collage.
Pictionary Desriptives	The transfer of what early humans saw visually in their topographies to that of the development of language. Another name for Birtur Linguistics.
Progressive Environmentalism	
Project Comparative Chart	The world is changing via DNA
real consciousness	The notion that enlightenment will bring forth progress.
Rule of Balance of Our Enviornments	A theory that supports the plus or minus of any population more in reference to "the stagnant period of time"
Russian Doll Analogy	That hominids were incrementally built over the course of two millions years versus what was historically thought as different species in different locations.
Saturation Point Theory	The saturation point theory is a process in evolution, it is an imaginery line that divides the first homo sapiens born and various other humanoid types still in existence. modify

	Followed by intermixing and reaching a uniform equilibrium of physical features over extended periods of time. No different than mixing several chemicals in one glass to create an end-product.
Seesaw effect on Skin	The impact of skin color of ices melting and heat rising by the belt of the equator.
Seven billion -click syndrome	
Shamanic date Era	Human and human proto-types who roamed from 1.5 million years to 10,000 BC
Shamanism	Inclusive of hominids are closest ancestry of our talking forefathers and their lifestyles.
Social Patenting	The generational patenting or identity of a social group of people to their habitats. As quick as 300 years upto 10,000 years the results are the same. Imagine a patent but this time socially done when no technology or outside intrefrence to create mickming really existed.
Stonehenge	Mention many location etc.
The Stagnant Period	Is the period of time classification when scientist deemed the approximate number of humans on earth was a relatively stable figure at around 55,000. It is known as the long term "crucial period" for the development of linguistic markings, better

	defined as regional names. Names given by our nomadic ancestors.
Time	A measurement or component to calculate or classify events. A measurement not really understood by humans.
Transitional Period (Also known as the era of Individualization)	The period approximately 100- 150 years ago with the start of urbanization. When individualization was promoted and people were forced to pick up their own last names. Urbanization resulted in the breakup of traditional living patterns and groupings of people globally that use to exist.
Turkic indigenous / Turkic Central Asian versus Turkish Anatolian Language category	An Altaic language belonging to the agglutinative family. Turkic linguistics belonging to Central Asian Tribes Modern Turkish, modified extensively in 1929 by Ataturk belongs to modern day Turks in the Anatolian region.
Turkic Language Categories and its multi-lateral variations	1. Birtur (Pictionary) / Early Humans -when the world had no written. Period *1,000,000 BC -15,000 BC* ➤ *Movement of indigenous people*

	crossing the Bering Strait to the Americas. See Time Travel Columbus mammoth case study
	2. Proto-Turkic Urheimat more associated with temporary settlements in other regions Period *15,000 BC- onwards.*
	The commencement of written and modification of words to include non-descriptive, but general labelling. See Nomadic Patterns.
	3. Turkic belonging to the Indigenous of Central Asia / close to a 100 or more complex variations of dialects. 35,000 BC *- onwards look at han dynasties of china.*
	4. Modern Turkish and the incorporation of European words into the language, Anatolian region. *1929- onwards*
	5. Making uniform the language across the board via Media tools. *Present day.*
	* Please note these are all very different categories and periods in time, as illustrated. They also belonged to the various religious groups of earth, with different customs, physical features, and dialects.
Unwritten portion of our Shamanic history	A period of time when humans had dialogues and no written account.

The Bering Confusion

"We are livid" they quietly said. We are a multicultural country; we have encountered a problem. Blankly they quietly looked and asked each other:

Are first clans of Earth, Asians? How can they be if these descriptive are a form of proto- Turkic from the many regions?

Are first clans of Earth black? How can they be if there is a standardization about them and they were found globally? Color is predominantly climatically based? She is right, but why didn't we think of that?

Could Natives just be Chinese now? But they have phonetics even belonging to Europe, and Africa?

In addition, there are so many pro-types, we can't visualize these people crossing. Are we ignorant?

First clans are Turkish? That is what they are now all saying. Canada is ruined they fumed!

We are all Turks? We are so damn confused! We can't be link to them. The Natives are now

Who released this information? We will by-pass her, and release this our own way they privately fumed!

§ God in this church I pray to you, injured is my soul, lost is my dominant arm, bored with tears I am.

Many days and nights at home alone, I read my hobby work to you. At home my papa teaches me Ottoman too.

Then in the middle of the night, like a lightening strike that came bolting down. A flash I saw.

A glimpse of our unified human past. Hot water came boiling down. Helplessness was the birth of the story that developed.

Frantic to those who protect I ran? Do you not understand social development; do you wish to ruin culture? Even our Natives are not shaman but a distinct culture.

You are the highest of the light? Release this information immediately, I yelled.

Then it all began, callous nights of house arrest, genetic tests, the violence ensued. Call it the "Bering Strait confusion."

A Canadian Heritage Poem

Kanada Tribe, Is This All Your Home?

We land on this land,

ice everywhere, it is 1534.

Do you think they will like us?

Indian what is your name?

They don't answer, keep trying,

they still don't answer.

I tell them mine.

Europe told us they are the Kana people.

I think they are from Kanada tribe.

Indian where is your home?

Pretend to do sleeping.

The land is cold,

our men are ghastly sick.

We need food,

danger lurks around us.

They point and say, Kanat, "Kanat'ta"

It looks like a small settlement.

We travel up north.

There are a hundred more,

Indian where are all of you from?

Hochelega, Iroquois, Oneida.

We must make way.

Europe is famished;

our people are hit with a plague,

we are so cold.

Our history says ugly,

we too have tribes and protect our own.

But tomorrow when we get stronger,

we will make way.

ABOUT THE AUTHOR

Born in Zeynep Kamil Hospital in Istanbul, Turkey and of Canadian heritage since the age of one, Yildiz Ilkin has presented this work for pushing social betterment by gluing her years of hobby research on many various academic subjects.

All which are and will be presented under the label of progressive environmentalism.

The following poems were done under grueling house confinement and relatively quickly. The objective being to convince with legal might that we have not reach our better evolutionary plateau. Therefore, if there are there are any comments, questions, historical edits etc. you can email her at ventingroom@yahoo.com.

Please note. The study notes for it can be found in her research center on www.starilkin.com at a later date.

Other Pages Of Interest www.vancats.com

www.oceanus12.wix.com/thefuture

www.ingramcontent.com/pod-product-compliance
Lightning Source LLC
Chambersburg PA
CBHW080515220326
41599CB00032B/6090